THE ENGINEERING IS EASY

Memoir of a Project Manager

THE ENGINEERING IS EASY

Memoir of a Project Manager

Bruce E. Podwal, PE, F.ASCE

© 2018 Bruce E. Podwal All rights reserved. No parts of this book may be reproduced or transmitted in any form or by any means, electronically or mechanical, including photocopying, recording or by any information storage or retrieval system without permission from the author, except for the quotations in a review.

This book is based, in part, upon composites of actual persons, events and companies where the characters, incidents and companies portrayed and the names used herein are fictitious. Any similarity of those composite characters, incidents and companies to the same attributes or actual background of any actual person or to any actual event or any existing company is entirely coincidental and unintentional.

All royalties are allotted to the City College of New York, Department of Civil Engineering
Printed in the United States of America
Published by Bruce E. Podwal
ISBNs 978-0-578-42944-1 (paperback), 978-0-578-42947-2 (ebook)

First Edition
10 9 8 7 6 5 4 3 2

Cover design by Marcus Klim

Study the past if you would define the future

Attributed to Confucius

The engineering is easy

Cited by A. B. Jenny

CONTENTS

Chapter

	Preamble	ix
1	Civil Engineering Student at CCNY	1
2	Junior Engineer	7
3	Highway Engineer	14
4	Squad Leader	19
5	Project Administrator	22
6	Project Manager	25
7	Managing a Multi-Disciplinary Project	37
8	Managing Multiple Multi-Disciplinary Projects	47
9	Managing Projects for Private Sector Clients	62
10	Managing a New York City Mega-Project	70
11	More Project Management Assignments	78
12	Managing an Asset Management Project in New York City	86
13	Managing Operations	91
14	Project Reviews	109
15	International Efforts	122
16	Managing a Design-Build Project in Turkey (With a Contractor as the Client)	131
17	Managing a Project in Hong Kong	153
18	Managing a Program in Guam	159

19	Managing a Public-Private Partnership Project in San Diego	165
20	Miscellaneous Responsibilities	173
21	Two Not So Successful Program Management Assignments	177
22	Program Manager of a Mega-Project in Houston	185
23	Firm-Wide Director of Projects	196
24	External Efforts	198
25	Proposal Management	204
26	Director on an International Engineering Company Board	208
27	Reflections	210

Appendix A	Project Management One-Liners	223
Appendix B	Good Practices/Lessons Learned	241

PREAMBLE

I had just finished leading a project management training session two decades ago, when a co-worker, Donald Pearson-Kirk, asked if I was going to write a book describing my project management approaches. While a book was more than I was ready to do then, I recalled what he said when I retired and decided writing my memoirs could be fun to do. Once I started writing about what happened, it became obvious to me that early in my career I presumed managing engineering projects would be simple, probably because I found most engineering tasks easy. Basically, I failed to grasp that management skills don't automatically flow from technical competence and had to learn the hard way that it's challenging to manage and lead a team to a successful project conclusion.

Further, as I memorialized self-doubts I had at various stages of my career, it caused me to wonder how someone like me happened to do so well in their career. I had neither apparent innate skills nor college training in management, leadership, communications or marketing. Yet, in spite of those weaknesses, I somehow found a path to success as a manager. It occurred to me that many of today's engineers must face the same hurdles I had, with their speaking, writing, management, and leadership skills paling when compared with technical skills.

At that point, practical purposes for my writing took shape: It would be a self-help guide, presented in a memoir format. I would use case studies to depict how I came to manage major engineering programs and serve on the Board of Directors of a major international engineering firm even though nothing in my youth and little in my college training prepared me for that future. The book would describe good and bad decisions I made over six decades as I progressed from an engineering student to Professional Engineer to Project Manager. It also would codify my experiences to inspire others to consider a career in engineering project and program management and to provide tips to help them succeed.

Although I understand what worked well for me may not fit everyone else's style, I hope this memoir-cum-self-help guide has relevance for those in project management and those supervising Project Managers. Moreover, while I use engineering projects to make a point, non-engineers should appreciate any activity with a scope, schedule and budget can be considered a project, such as running a restaurant, marketing a product to the public, or planning a wedding. Applying the good practices mentioned in this memoir means it's more likely both engineering and non-engineering projects will succeed.

It's worthwhile to clarify the terms project and Project Manager and program and Program Manager used in this memoir.

- **Project**: Assume an owner wants to widen a two-lane highway to four lanes for two miles. That effort is a "project." In general, a project has a contract that defines the services (often referred to as the "scope") to be provided by a date (the schedule) for a cost, fee or budget.
- **Project Manager:** Each of the various firms and government agencies involved in the widening would have a "Project Manager" (or some similar title) to manage that firm's scope, schedule and budget on the project. The primary firm on the project may have several subconsultants each performing a discrete project task, say the geotechnical engineering or the lighting design, with each subconsultant having its own Project Manager.
- **Program**: If the client wants to widen a highway for, say, 20 miles subdivided into several projects, the overall effort and coordination has become much more complex and is a major project that could be referred to as a "program."
- **Program Manager**: A "Program Manager" manages programs and occasionally that role may include non-traditional engineering efforts such as obtaining funding to finance the program.

Memoir chapters are by topic and not chronological. Initial chapters discuss what I did to become a Project Manager and describe the skills necessary to manage basic projects; those chapters would be very useful for those considering project management as a career or managing their first projects, as well as for their supervisors. Case studies in the later chapters discuss the additional leadership skills required to manage complex projects or programs, including multi-disciplinary, international, design-build, and public-private partnership projects and are intended for more experienced project managers and upper management. Strategies to avoid potential pitfalls are described throughout the book.

General Omar Bradley wrote, "I learned that good judgment comes from experience and that experience grows out of mistakes."
As I memorialized anecdotes, I was reminded when, how and why I made mistakes. These mistakes ranged over such varied circumstances as inadequately dealing with co-workers with potential bias and sexual harassment tendencies, failing to provide proper oversight of project activities, and not realizing when I lacked the skills to accomplish my assignment. I've attempted in this book to show how I had learned from my past mistakes the next time I faced a similar situation. My hope is this book will help others accept they'll make mistakes and realize they must make sure not to repeat them. To that end, Appendix A lists numerous cues I relied upon in my career to remind me how to avoid missteps. The cues are catch phrases that I refer to as "one-liners." Further, throughout the memoir, I've discussed both good practices and lessons learned and the context of their application. These good practices and lessons learned are organized by relevant chapter in Appendix B[1].

I studied engineering because the idea of traveling around the world to build bridges, tunnels, and other engineering projects excited a 13-year old from the Bronx. Along the way, my career evolved from design into project management, and after managing large, diverse and complex projects, I managed programs of multiple projects. At one point in my career, I left project management and became an operational manager overseeing a region of twelve offices in seven states. However, my love for projects stayed strong as I found operations management somewhat tedious. I longed for the challenge of bringing projects to a successful conclusion. The pride and satisfaction of seeing the public use those projects were other reasons why I moved back to managing projects.

---

[1] A superscript shown as (1-1) refers to Chapter and Number of a good practice/lesson learned that is listed in Appendix B.

There are those who think, rather than performing hands-on project management, the only way to be successful is by managing operations (e.g., running an office or a company). Not so; by managing projects, I was amply rewarded in my professional career, highlighted by being elected to the Parsons Brinckerhoff[2] board by my co-workers and to the Board of Direction for the American Society of Civil Engineers. Another counter argument is what Marty Rubin (then Parsons Brinckerhoff's Manager of Western Domestic Operations) did when pursuing a large program management assignment for Los Angeles County Metropolitan Transportation Authority. He told the client he would give up his role managing western operations if the firm won the assignment. And when the firm won, Rubin relinquished his operations role, supporting the view that project management is at least as important to an organization as regional or firm-wide leadership.

Many people moan and groan about how they don't enjoy their job. I tell them that's why it's called "work;" otherwise, it would be labeled "fun." As for me, I enjoyed being an engineer, which I always felt had purpose and meaning by providing lasting benefits to society. Specifically, I'm proud anytime I can tell those with me that I helped make this road on which we're riding a reality. A major reason I enjoyed my career were the people I worked with, especially those at Parsons Brinckerhoff for the fifty years I was there. That firm's culture attracted ethical individuals who were highly qualified in their field and a pleasure to work alongside. While there always were days I wish never happened, by and large I was fortunate someone thought what I did was valuable enough that they paid me to do what I enjoyed.

As a witness to over a half-century of changes and progress in the engineering profession, I've seen many advances including the evolution from the use of slide rules and log tables to computers and the maturation of engineering sub-disciplines such as geotechnical and traffic engineering. I've watched many small, local firms consolidate over those decades into internationally-focused, publically traded mega-conglomerates. Further, when I began as an engineer, such words and phrases as diversity,

---

[2] Subsequently acquired by WSP.

affirmative action, sustainability, infrastructure, environmental impact and occupational safety didn't exist or had different meanings than they do today. Historically then, this memoir also provides a brief look into the vanished world of engineering that existed 60 years ago.

In the 1970s comic strip "Pogo," the strip's characters reading a memoir said, "We know what happened wasn't anything like that, as we were there."
The character who authored the memoir replied, "Are you going to believe your faulty memory or what's down in black and white?"
It was Will Rogers who described a memoir as when you put down the good things you ought to have done and leave out the bad ones you did. That said, of course this memoir is subject to the vagaries of my memory, although I hope it conforms to what others remember as well. If not, then please allow me a little latitude. And, Dear Reader, if I put someone's words in quotes, accept that my memory is paraphrasing what I recall was said.

I acknowledge Carl Selinger, who gave me many useful tips on what to include in a memoir so it would be introspective rather than simply autobiographical. Based on Chuck Fuhs' comments, I made a greater effort to explain the actions I took. Chuck is one of those rare individuals whom everyone calls a friend. Bruce Fuller's suggestions significantly improved the overall flow of information and focus of this book. I'm grateful to my daughter, Stef Woods, for her edits and recommendations that added much needed clarity. Mary Karr's *The Art of Memoir* provided numerous ways to improve content. Many thanks to Gilda Barabino, Greg Kelly and Bob Prieto for their support and suggestions. I also thank Marcus Klim for his creative design of the book's cover.

This is not an autobiography so my personal and family activities are not mentioned, except as they directly relate to my career. I'd be remiss, however, to fail to note that Mary Woods, my late wife, was a major influence in helping me develop the people skills required by a manager, and I dedicate this book to her memory.

CHAPTER 1

CIVIL ENGINEERING STUDENT AT CCNY

"A Civil Engineer, what's that?"
I was 13 when Arthur Goulet, a friend who was a grade ahead of me at DeWitt Clinton High School in the Bronx, New York, said he was going to be a Civil Engineer. It sounded fascinating when he described what a Civil Engineer did. Born late in the depression in 1938, my career goal was "get a job" — to stop being a financial drain on my parents. Moreover, it was a time when it seemed there always were jobs available for engineers, which was very important to me. When I asked what skills were needed to become a Civil Engineer, Goulet said math skills were important, and that clinched it for me, as math was my strength. Had he said an engineer should have a general understanding of how things worked and how to tinker with machinery, I would have looked for another career choice. I never had those skills as a youth, probably because my parents weren't handy (the only tools we had at home were a hammer and a flat-head screwdriver). My father taught me how to change an electric plug, but little else technically.

I passed the entrance exam and was accepted at 16 into the City College of New York ("CCNY") School of Technology in 1955. CCNY was tuition-free in those days, and I only had to pay for books, student activity fees and lab fees. As I'd won a New York State scholarship, my undergraduate education cost very little. Back then, CCNY was a commuter college with no dorms, so I lived at home further saving money. Also, I decided to complete the four-and-a-half-year curriculum in four years, to be able to start earning a living as quickly as possible. Without asking for approval, I signed up for more credits per semester than I should have been allowed to take. I didn't apply for ROTC as that would have kept me from graduating in four years, although I expected to be drafted sometime after graduation.

Not surprisingly, I struggled a bit academically because of all the extra credits I took each semester. Fortunately, I passed all my classes and stayed on track, though I did worry a few times if I had made the right call to accelerate my studies. By taking extra courses per semester and night courses during two summers, I was able to graduate in June 1959.

I still recall tips from some of my civil engineering professors. Professor Bernard Kaplan explained the merits of prior planning before commencing design, while Professor David Muss stressed the value of having an executive summary at the front of a report.[1-1] Professor Donald Brandt showed how to criticize the action but not the person, when he'd say, "That's a way," to a student's flawed approach to a problem, before describing the proper approach.[1-2]

And Professor William Brotherton's response, "The better one," to the question, "Which officer salutes first when two officers of the same rank meet for the first time?" taught me that it's more important to be courteous than egotistical.[1-3]

While these tips were not directly part of the curriculum, remembering them eventually helped me become a better manager and leader.

As a sophomore, I took a surveying course where I was part of a three-person crew, with Archie Zosuls and Vaino Ader, surveying portions of Van Cortlandt Park in the Bronx. The surveying instruments we used were appropriate for 1957 and lacked electronic distance measurement capabilities as equipment with such features was still a few years away. Zosuls had worked as an intern the previous summer for DeLeuw, Cather & Brill ("DC&B"), a consulting engineering firm, and suggested Ader and I apply for jobs the coming summer. Zosuls told us that when he interviewed for the job, things weren't going too well when the interviewer, call him Ben[3], pulled out a cigarette. Zosuls said he quickly got out his lighter and lit Ben's cigarette. After that, he felt the interview went much better, and he ultimately was offered a summer position.

As intern for an engineering firm sounded much better than office boy for a publisher, my job the previous summer, I applied for a summer position with DC&B. The interview sequence went exactly as Zosuls said it had for him. After Ben and I talked a while, Ben took out a cigarette. I pulled out the lighter I was carrying, even though I didn't smoke, and lit his

---

[3] Proper names are fictitious when preceded by call it, call him or call her.

cigarette.[1-4] And the following week, Ben offered me a summer position in their Buffalo office. In my application, I asked for $60 a week and was very surprised when they offered me $76 a week (which was their going rate for an engineering intern with two years of college). I naturally accepted it.

Summer Intern:

After I was hired, Zosuls and Ader also applied to DC&B and were hired to work in Buffalo for the summer. We roomed together in an apartment a few miles from the office. I was assigned as an Engineering Aide in the Highway Department, where I worked on a project to improve Hyde Park Boulevard in Niagara Falls, NY. I performed routine tasks, such as tracing plans and performing basic calculations. Nevertheless, it was exciting to think of how my simple efforts would help complete the road improvements. I worked primarily on typical sections[4]. One message stayed with me throughout my career; namely, if the typical sections don't make sense, it's likely the design is flawed. Years later, whenever I reviewed a set of highway plans someone else prepared, I started by looking at the typical sections. The preceding is an example of seeking key indicators to assess if the person who did some work understands what they're doing.[1-5]

Half my summer was spent reviewing cross-section sheets[5] to estimate the volume of earthwork to be removed or added. To sum the earthwork quantities from hundreds of cross sections, I had to add a long, hand-written column of six-figured numbers several times a day. It was pre-computer days, and additions were performed by typing numbers into a hand-cranked, mechanical adding machine, which printed the typed numbers and their total onto a tape. After the machine totaled the inputted numbers, I would back-check the numbers on the tape against

---

[4] A typical section is the basic shape of the road perpendicular to the direction of travel depicting roadway elements such as the number of lanes and their widths and the pavement materials and their thicknesses.
[5] A cross section depicts a slice through a proposed road and the existing ground perpendicular to the direction of travel.

the numbers in the hand-written column to make sure I had input the correct numbers. I quickly learned it was almost impossible to input so many numbers into the adding machine without making an error. Only after careful back checking, could I be satisfied I had typed everything correctly and the total was accurate. Numbingly boring to do day after day, it was at that point the idea for a practical joke hit me.

First, I carefully added one of the hand-written columns of about 30 six-digit numbers and wrote the total on a piece of paper on my desk, which I could see, but where it wouldn't be obvious to others. Next, I asked the 16-year old office boy, call him Ken, to add the column of numbers using the adding machine. Ken dashed off and soon returned with the total. As I anticipated, Ken's haste and inexperience resulted in input errors and his total was wrong. I took the original hand-written column of numbers and ran my finger down the 30 numbers in two seconds. After I peeked at the total written on the paper, I told Ken his total was wrong and gave him the correct answer. I then told him to redo the addition more carefully. Ken was stunned! He couldn't fathom how in two seconds without an adding machine I could believe my answer was correct and his wasn't.

He went away to redo the addition and, when he returned, his answer now agreed with mine. "How did you add those numbers so quickly?" Ken asked.

I replied, "By adding all six digits at the same time rather than adding columns one digit at a time."

I suggested he go home and practice adding columns of two-digit numbers in his head until he had that down pat and then expand to three, four, five and finally six-digit numbers. To this day, I wonder if Ken tells of meeting someone 60 years ago who could add long columns of numbers in a flash.

One afternoon we were hand coloring several sets of plans that had to be delivered to the client at the end of the day. All the summer interns and drafters were pressed into service to color the plans. We were working as quickly as we could when I noticed the head of the office, Harry Quinn, had picked up some colored pencils and was helping us finish the task. My respect for Quinn increased, as he showed that no one is so important that they can't pitch in to help at the most elementary effort when necessary.[1-6]

After two months in the Buffalo design office, I was reassigned an hour away to a DC&B field trailer in Dunkirk, New York for a month. While working there, I lived in a rooming house in nearby Fredonia. My main task was helping finalize payment quantities on a rural section of the Erie Thruway where DC&B was inspecting construction. Co-workers at the trailer included the Resident Engineer and Chief of Survey, both of whom were graduate Civil Engineers, while the balance of the field crew were locals DC&B hired from a nearby tavern. I was self-conscious to learn I was making more than the junior inspectors who had been on the project for over a year. Even though they had no formal engineering training, their years of on-the-job experience meant the junior inspectors were much more knowledgeable about inspecting construction than I was. It occurred to me that when two people compared the salaries each was earning, it was likely one would become resentful. Thus, for the rest of my career, I avoided telling others how much I was being compensated.

The road opened to traffic while I was in Dunkirk, and a highlight was when Governor Averell Harriman rode by in his limousine on opening day and waved towards me. I impressed easily in those days. Even though I was involved in my career on many construction inspection and construction management projects, the month I spent in Dunkirk would be the only time I ever worked full-time at a construction site. I'm sure a more balanced field-office ratio in my career would have given me a broader understanding of all aspects of highway engineering. However, I really enjoyed office work over field and never sought a different balance over time. In general, I noted most American Civil Engineers prefer one over the other, and it's relatively few who don't care if it's office work or field work that they do.

That summer solidified my desire to be a Civil Engineer. The work was interesting, and because I wasn't mechanically inclined, I recognized other engineering disciplines (such as mechanical and electrical) weren't for me. I sensed it would feed my ego when family and friends would be as impressed as I was that the civil engineering projects I would work on had physical presence and obvious value. I also settled on highway engineering as my preferred specialty, because I felt a Highway Engineer had a broad role on projects. Structural Engineers obviously focused mainly on the structures (the trees), while the Highway Engineer's expansive role (the forest) appealed more to me.

CHAPTER 2

JUNIOR ENGINEER

As graduation from CCNY approached, I applied for positions with DC&B and the New York State Department of Public Works (the precursor of the New York State Department of Transportation or "NYSDOT"). I received offers from both, and decided to accept the one from DC&B to work in their New York City office. The decision to work in the private sector over the public sector primarily was based on enjoying the three months I worked for DC&B.

I was assigned to the Highway Department because I had been in that department in Buffalo and happily started down the road [*pun intended*] to become a Highway Engineer. My starting salary of almost $5000 a year was slightly above the average of what my classmates were offered for their first job, most of whom went to the public sector.

My first project was I-95 in Florida; our task was to develop the proposed highway's alignment through Miami, where I performed preliminary alignment designs after a senior Highway Engineer (call him Herb) laid out the base alignment. I decided to study what Herb was doing so I could learn how to lay out a base alignment myself. In doing so, I noted the highway's north-south alignment appeared to wiggle east and west unnecessarily. On closer inspection, I saw the planned route seemed to require demolishing every Catholic Church in the corridor. It bothered me to think a co-worker could be prejudiced. I realized the implication was serious, but felt I was too junior to question Herb why he set the alignment that way. Neither did I voice concerns to anyone else, as I wasn't confident the opinions of a 20-year old junior engineer with one week's experience had significance against a senior engineer.

To this day, I wonder about Herb's motivation. Was he anti-Catholic? Was he pro-Catholic and knew the Diocese preferred that the State acquire no longer needed church property? Was he a good engineer doing his job properly? Over the years, I became an adherent of Occam's razor and normally assert the simplest assumption is better than a complex one. In this case, it would mean Herb designed the alignment properly and just happened to hit all those churches. I hope most junior engineers, in the same situation today, would do better than I did; that they would overcome any reluctance and speak up or call their firm's whistle-blower hotline and let someone more senior determine the real reason.[2-1]

Working in the private sector typically has a lower degree of security than in the public sector. Such jeopardy was borne out when there was a massive layoff at DC&B on a Friday afternoon in my first month. I sat at my drafting table as P.G. Dover, the Chief Engineer, fired, among others, the engineer to the left of me and then the engineer to my right. Obviously, Dover was carrying out the wishes of the firm's owners to cut costs by reducing staff. It was easy to follow Dover as he walked to the desk or drafting table of the next person he told to go, because (except for upper management) no one had a cubicle or office, and there were no walls around anyone's work space. Everyone sat frozen at their work area watching the scenario play out and worrying they might be the next one fired. The situation continued for an hour as Dover laid off employee after employee. As the least experienced engineer in the office, I kept expecting Dover would dismiss me next, but when the firing ended, I was still on my drafting stool.

I went home relieved, but wondering why I survived over more experienced engineers who clearly knew more than I did. A few days later, the reason I wasn't fired dawned on me — I was making at least $4 a week less than every other engineer in the office and could do routine work more cheaply than anyone else could. Still, I remained concerned for the next several weeks that another layoff might be coming and I wouldn't survive that one. As time went by, it became apparent the situation had stabilized, and I was able to relax.

Messrs. DeLeuw and Cather were based in Chicago, while Mr. Brill was based in NYC. The DC&B partnership split not long after the layoffs, and Brill began running the reorganized firm, now called Brill Engineering Corporation ("BEC"). At BEC, highway engineering encompassed several areas that today many firms treat as separate disciplines, such as traffic engineering, drainage engineering, cost estimating, and specification writing. I had heard some firms stuck you in one area (for example, you designed only drainage), and I believed becoming a specialist wasn't the best way to spend my early years in civil engineering.

I was fortunate I never was pigeonholed in one narrow discipline area and could learn them all. I did horizontal and vertical alignment designs, prepared traffic signal plans, worked on interchange designs, developed drainage plans, prepared construction cost estimates, and wrote specifications. It was true on-the-job training, as I was assigned tasks in which I received little instruction and guidance. I found some pamphlets on how to do highway geometric design, which helped me avoid foolish mistakes. Most likely most of my work products back then were adequate, which seemed to meet the expectations that BEC's senior managers wanted — an acceptable product at an economical price. At the time, I sensed that with more guidance and training, I could do a better job. For example, years later, an excellent specifications writer at Parsons Brinckerhoff, Joe Goldbloom, told me it was important to write specifications so both the contractor could bid and build the item and an inspector could determine if the work was constructed as intended.[2-2] Had I had that advice when I worked at BEC, I'm certain I would have written clearer specifications.

BEC's office was basic. There were large fans everywhere, and windows were open during the summers because there was no a/c; you brushed city soot from the drawings on your desk every day. There was a single phone on the wall for all non-management employees to use, as keeping costs low was critical (a competitor had a payphone for its employees). We used pencil extenders to get extra usage out of pencils as a cost-saving approach. A principal of the firm, not wanting to pay for all that extra electricity, threatened to fire the person who bought a 100W bulb and replaced a 60W bulb in the ceiling light fixture over his desk to enhance visibility.

Computer aided drafting was over a decade away, and engineers did designs in pencil on paper, which they turned over to drafters to trace in ink onto the final plans. Some drafters had engineering or technical training while others were graphic designers with good drawing skills. It was easy to tell which drafter had worked on a plan sheet by the handwriting style. A few drafters had so much pride in their work, they wouldn't work on a plan sheet another drafter started. One drafter told me that approximately 10% of men are partially colorblind, and we must be cautious when drafting plans, charts and exhibits in color. Everyone owned their own tools such as scales, triangles, curves and straight edges. About once a quarter, a traveling salesman came around carrying suitcases filled with tools for us to purchase.

There was a BEC drafter who was a bit of a jokester, reminding me of my college days in Mechanical Drawing class when I'd put a famous date on drawings, such as July 4, 1776, rather than the current date. One of the drafter's pranks was adding three small drainage ponds in the shape of his initials when drafting a topographic map. Another prank was to mislabel things — a horseshoe pit in a recreation area became a horsesh#t pit. A client caught that last one, and BEC's Project Manager had to talk our way out of that embarrassment. Even though I see humor in many things, that incident taught me some clients take sophomoric pranks seriously, and I must be careful when checking everyone's work.

Graduate School:

While writing this memoir, I read a *New York Times* article by Tom Friedman, where he said going to college for four years and then spending that knowledge for the next 30 is over — to be a lifelong employee anywhere today, you have to be a lifelong learner.

Becoming a lifelong employee just wasn't something I thought about three months after receiving my bachelor's degree. While I hadn't planned to go to graduate school full-time for financial reasons (I needed a regular paycheck), I knew it would be foolish not to take advantage of the fact that BEC paid the tuition for graduate courses. I also felt taking courses would satisfy my curiosity and interest in learning about different things.

For those reasons, I began taking two evening courses a semester from NYU Polytechnic School of Engineering (then the Polytechnic Institute of Brooklyn or "Poly"), until I received a Master of Science in Civil

Engineering in 1963. I didn't take courses during the summer, so those months were like a vacation because I only had to work 40 hours a week.

My CCNY undergraduate courses had done an exceptional job preparing me for Poly's graduate level. I didn't concentrate on highway engineering courses at Poly as I believed becoming a generalist would benefit me more than focusing on one area. Instead, I took courses in structural engineering, foundation design, airport engineering, traffic engineering, hydraulics engineering, economics, probability & statistics, and city planning. Such a general approach is not likely to be permitted today, as a Master's degree now typically requires specialization. Back then, a Master's degree at Poly required attending a non-credit seminar series where guest speakers came to speak about different topics. These seminar sessions were very helpful in civil engineering areas where my understanding was relatively weak, generally disciplines other than highway engineering.

Another Poly Master's degree requirement involved completing a major report or a thesis. My report was on "Comprehensive Parking Studies," and my advisor was Professor Lou Pignataro. At the time, he was writing possibly the first textbook on traffic engineering, and some Traffic Engineers of my generation consider him the father of that discipline. Before this textbook, and computer modeling a few years later, traffic engineering often was educated judgement. An Albany-based, NYSDOT traffic engineer had a Ouija board in his desk. He took the board out when asked a question such as, "How many lanes are required to handle anticipated traffic 20 years in the future," on a specific highway ramp under design. After pretending to fiddle with the board, he would pronounce something such as, "That ramp should be two lanes wide."

Preparing my Master's report helped me understand how planning, design, construction, operations and maintenance influence one other. I also learned how to organize a technical report and how to defend my work before a review committee. The skills I gained preparing the report benefited me greatly over the years.

A graduate degree helped me in many other ways. I improved technically by taking courses covering topics related to what I worked on in the office. Also, the graduate coursework proved useful when I took the test a few years later to become a Professional Engineer. Further, a graduate degree gave me added stature over those lacking a Masters and aided my advancing more quickly than others. Today, I would say it's even more important for engineers to have a graduate degree, especially given that many of today's undergraduate engineering programs involve fewer credits than in my day, which was 145 credits at CCNY. Tom Friedman's article reinforces why lifelong learning really isn't optional.

Time to Move on:

When I started at BEC after graduation, the firm had no computers, and we performed calculations using mechanical calculating machines and slide rules. The department had two mechanical calculators, each on its own work-station desk. You used whichever mechanical calculator was free, whenever calculations by pencil and paper or slide rule were impractical. IBM contacted the firm's management in 1960 and invited one Highway and two Structural Engineers to visit their facilities to see how computers could be useful. The Structural Engineers came back from the visit and said their department would benefit by having a computer and training staff to use it. The Highway Engineer (who had 30 years' experience) said he could do everything a computer did with a mechanical calculator and recommended against the Highway Department buying a computer. Such shortsightedness was why I began to feel BEC's senior management was old and tired and lacked the vision necessary for the firm to prosper. In fact, I guessed Brill was only interested in finding his personal retirement exit strategy and wasn't concerned about the long-term future growth of the firm and its employees.

Deciding there was no future at BEC, I started looking for a new job in March 1961. Co-workers had said New York City's consulting firms had a gentlemen's agreement that they wouldn't hire from each other as a way to keep salaries low. It now would be considered restraint of trade or something similar. When a firm had an opening, they posted the position with an employment agency, rather than placing their own advertisement in the classifieds. I saw a *New York Times* ad by an employment agency for a Highway Engineer with two to five years' experience, to work in a NYC office for considerably more than I was making at BEC. The agency wouldn't tell the name of the hiring firm until I signed an agreement that I would pay the agency two weeks' salary if hired by that firm.

After signing the agreement, I interviewed for the position at Parsons, Brinckerhoff, Quade & Douglas with Dick Duttenhoeffer, Deputy Head of the Highway Department. Duttenhoeffer (who would become a partner of the firm) asked me a series of questions he gave all applicants on different areas of highway design. Because I hadn't specialized in one area, I was familiar with them all and nailed every question. I left the interview feeling optimistic and, sure enough, was offered a position a few days later. As I was two months' short of two years' experience, Duttenhoeffer offered me a few dollars less per week than was mentioned in the ad — still a nice salary increase for me. I accepted with the proviso he'd review my performance in two months, and my compensation adjusted accordingly. Given how well I did at the interview, I was confident I would get an increase. When Duttenhoeffer reviewed my efforts two months later, he not only raised my pay, but gave me more than the minimum in the ad. In all the years we worked together, he always treated me fairly. Duttenhoeffer was a World War II veteran who worked as a draftsman after the war, while getting an engineering degree by going to college at night on the GI Bill. He was perhaps the most meticulous Highway Engineer with whom I ever worked.

I'd heard you could do well financially by moving every two years, and it proved true in this case. I already was assuming I would stay at Parsons Brinckerhoff for two years, before looking around for my next job. However, while I had heard of Parsons Brinckerhoff, I knew little about the firm's culture or values. I didn't yet realize how fortunate I was to be employed by that firm and how two years could stretch into five decades. For the first two years at Parsons Brinckerhoff, I answered a few ads for positions elsewhere. Luckily, none panned out before I learned enough about the firm to know I should stay.

CHAPTER 3

HIGHWAY ENGINEER

At Parsons Brinckerhoff, I started as a Junior Highway Engineer (and soon after became a Highway Engineer), where I was assigned on a project-by-project basis to various Squad Leaders in the Highway Department. Usually, the most senior Highway Engineer working on a project was the Squad Leader and the other engineers on the project reported to him (Squad Leaders were all male back then). Squad Leaders were barely entry-level managers as they didn't participate in evaluating the performance of those reporting to them. In some other firms, a Squad Leader was called a Job Engineer or Project Engineer.

The need for engineers to design highways had exploded in the late fifties/early sixties once the Interstate Highway Program got into full swing. As there weren't enough engineers coming out of college to meet the demand, almost half the designers in the firm's Highway Department hadn't graduated college. Many were veterans of World War II or the Korean War who had gone to college on the G.I. Bill, but dropped out after a year or two. Some of those who hadn't graduated were nevertheless quite talented and rose to become Squad Leaders and even Project Managers.

One of my first assignments was on the final design for I-80/95 just west of the George Washington Bridge in northern New Jersey. Twenty years later, it was slightly depressing when it became the first project I worked on that became so outmoded it was reconstructed to handle the latest highway standards and traffic demands. Starting in my early years at Parsons Brinckerhoff, I worked with some excellent managers and leaders. At first, still more concerned with having and keeping a job, I really wasn't thinking about becoming a manager myself. Nonetheless, I was noticing the managerial actions that generated positive outcomes and was filing them away in case I might use them myself someday.

I often worked for Marty Rubin, a Squad Leader and fellow CCNY alum, who would become a partner of the firm. Rubin had a great management style. Even when he chewed us out for doing something wrong, he came across as both firm and caring, and it made us want to work harder for him.[3-1] Rubin was Squad Leader on a project for the extension of Sunrise Highway in Bayshore, Long Island, NY, where my task was to design the roadway drainage. When I was about 85% complete, he asked me to go to the site to confirm the drainage design made sense. It was the first time I'd done something like that, and it was enlightening.

While walking the site, I noted what was depicted as existing on plans I was carrying wasn't always what existed in the real world. For example, some topographic features were not where they were shown on the plans. Fortuitously, it rained that day, and while it was difficult marking notes on wet plans with my eyeglasses covered by raindrops, the rain meant I could see the direction water was flowing and where water was ponding in unanticipated places. Each of those situations helped me determine if the existing contours depicted on the plans were correct. By the time I'd finished my site visit, the plans I'd brought were filled with annotations of all the changes that had to be made, and I now knew where the design had to be adjusted. After that experience, I knew to take what was shown as "existing" on plans with skepticism unless verified by a site visit.[3-2]

In 1963, Parsons Brinckerhoff was awarded a major assignment in San Francisco-Oakland on the Bay Area Rapid Transit project, and it became the firm's catalyst to both win more transit projects and become bi-coastal. Rubin was relocating to California to be Chief Civil Engineer of that project. He asked if I would join him there, and I said I would. However, Duttenhoeffer didn't approve of my relocation. He felt I should remain in the Highway Department in NYC, and Rubin should look to hire engineers locally in California. While I hadn't asked for the move to California in the first place, I was disappointed I wouldn't be going. It was the first of many times I agreed to relocate, although my first actual relocation wouldn't happen for another 25 years. The only time I didn't agree to a relocation was when no one could explain to me what I would be doing over the next year or so. Given the vagueness of their answer, I declined as I felt the disruption to my family didn't warrant agreeing to relocate on speculation it might be beneficial for my career.

George Oliger, a classmate from CCNY, worked at Parsons Brinckerhoff in a four-person group of drainage and geotechnical engineering specialists. In lieu of struggling on my own, I'd go to that group for assistance whenever I had a problem in one of those disciplines that no one in the Highway Department could solve. The senior drainage specialist, Enrique Aiseks, used the Socratic Method to help me. Aiseks never directly answered my questions, but would point me in the right direction, encouraging me to find the answer to my own problem. While it was often frustrating that he didn't just give me the answer, it helped me learn how to solve similar problems unaided in the future.

One September day, everyone started looking for a set of drawings that seemingly had disappeared. Someone recalled a summer intern had been working on the plans before returning to college, and we called him to see if he remembered where the drawings might be. It turned out he had them at home. On his last day at work, the intern had asked an engineer if he needed the drawings. The engineer, who didn't need the drawings <u>at that moment</u>, responded, "No."
The intern assumed that meant the engineer would never need the drawings again and took them home as a souvenir. It showed how a casually conveyed response to a question can come back and bite you.[3-3]

The intern's motive was benign compared with that of a senior engineer in California who reported to me many years later. The senior engineer had resigned from the firm and was going to work for a competitor. As he was leaving the office, we saw him taking a dozen boxes, which seemed excessive. I looked at some of the boxes and found he was taking proposals, marketing material, company manuals and other proprietary information. He was furious I checked his boxes, and said they were personal and we didn't have the right to look in the boxes. That stopped me for a moment — did I indeed have the right to check his boxes? To be sure, I checked with our in-house counsel who confirmed we had the right to inspect anything on our premises, and it was our call whether to release any material created by the firm. We then searched all the contents and allowed him to take only materials that weren't company property. I certainly would have cut him some slack had he only taken a few items he had worked on as souvenirs. However, as far as I was concerned, his audacity in taking so much proprietary material meant he deserved nothing other than personal items. For as long as I knew him after this incident, he was unrepentant about his actions. As there was no question he was taking company-owned material to use at his new firm in competition with us, I never understood why he didn't just accept the fact he was caught.

There was a small group of designers who worked on conceptual planning for transportation projects. The head of the group, Charlie Louis, often had clients wanting to see how things were going before he was ready to release plans for distribution. Prior to showing incomplete plans to clients, Louis marked those plans with a bold label: "WOIKOPI," which was "work copy" pronounced in an exaggerated Brooklyn accent. That way, if the client demanded a copy of the exhibited plans, the plans were already clearly labeled to show they were incomplete. Learning from Louis, I often did something to draft reports and plans that would reinforce the fact the document was not final. For example, I'd truncate a meaningless sentence in a report so anyone (client or supervisor) reading the report would realize it wasn't a final version.

I expected to be drafted into the military at some point and wasn't surprised when I received my draft notice during the Berlin Crisis in late 1961. However, Duttenhoeffer said my work on Interstate highways qualified me for a deferment, as those roads were part of the national defense network. My draft board agreed, deciding it was better for the country if I continued designing Interstate highways, and gave me a deferment which became permanent when I turned 26.

While it may be generational, throughout my career, I typically wore a suit and tie at work. And when Casual Fridays began many years later, I wore a sport jacket and tie and tried not to dress so informally that I looked sloppy. It seemed logical that clients would assume I didn't care about how their project looked if they thought I didn't care how I looked.[3-4]

While I was still a junior engineer, an engineer was hired (call her Martha) and became the only female engineer in the Highway Department. She was from Canada and had several years of experience there. One day, she told me that some engineers in the department were continuously picking on her and making unflattering comments about her and her skills simply because she was female. In those pre-feminist days, there was no one Martha could go to for help. She spoke to me as a friend seeking my moral support. I had overheard some of those misogynistic comments and knew she was telling the truth. Unfortunately, I was never more than a sympathetic shoulder, and probably even her supervisor, would have only told her to toughen up. Soon after, Martha left the firm. I felt badly because I realized I should have found a way to be more helpful.

There are obvious lessons to be learned here, although I expect a similar situation is less likely to happen in today's environment. First, today Martha probably would know to talk to a supportive Human Resources Department that exists in most modern firms. Second, some of her co-workers are more likely to speak up to say gender bias is uncalled for — I know I would.[3-5]

CHAPTER 4

SQUAD LEADER

By my third year at Parsons Brinckerhoff, I had ceased looking for a job outside the firm as I found the growth opportunities at the firm were good, the projects interesting and my co-workers always ready to lend a hand. The idea of moving to another firm for a few dollars more had become meaningless. Working with people I liked and for a company I respected, was a blueprint for happiness [*forgive the cliché and rather note the engineering analogy*]. After that point, whenever recruiters contacted me trying to entice me to work for another firm, I'd say I wasn't interested, although, I'd give them the name of anyone I heard was being laid-off.

It was important to become licensed as most senior engineers at Parsons Brinckerhoff were Professional Engineers. I had taken and passed the first two parts of the exam when I graduated from CCNY. I took the third and final part of the examination as soon as I was eligible (four years' experience and at least 25 years old) and became a licensed Professional Engineer ("PE") in New York. Over time, I became registered in seven additional states.

Duttenhoeffer appointed me Squad Leader after three years with the firm, even though I was only 25 with much less experience than other engineers who weren't promoted. I was surprised I advanced more quickly than those with more seniority, as I naively assumed people got ahead strictly on time-in-grade. After thinking about it for a while, I realized Duttenhoeffer must have looked at more than seniority in moving staff along; it's likely he took into consideration that I was a PE and had a master's degree when he decided to pick the next Squad Leader. I didn't appreciate it then, but I believe he also may have noted that when I did a work effort, I tended to visualize it as part of a project continuum and automatically planned for the next work effort while completing the current one.[4-1] After all, it's logical that those who only do their assignments and nothing more are going to fall behind those who are proactive in moving the work along to the next phase and preparing for the next step before being told to do so. As for me, I'm almost always thinking about what comes next, as I view it as an eventual time-saver.

I participated in social activities at the firm because they were fun, not appreciating they would help me bond and develop long-lasting relationships with co-workers. When Rubin left for California, I replaced him in a lunchtime bridge foursome. My bridge partner was another Highway Department Squad Leader, Howard Chaliff, with whom I also played on the company softball team and would work with later in California. I joined the company bowling league, where I was elected as league President. My bowling teammates included Yalcin Tarhan and Vijay Chandra, two excellent Structural Engineers with whom I worked on many projects, and John Peterson and Jimmie Bowen, who both worked many years with the firm in administrative roles. In addition to resulting in a more pleasant work environment, the social connections I made proved helpful to me over the years.[4-2]

When I began at Parsons Brinckerhoff, the firm was owned by six partners, one of whom was Bill Bruce, who oversaw the Construction Inspection Department. He told me about his initial job after graduation where he was to inspect construction of a steel building. When he first arrived at the site, he saw steel workers effortlessly running around the narrow steel beams several floors above ground. After Bruce climbed the ladders to the fourth floor to start inspecting rivets, every worker stopped to see if this rookie would walk the steel to check if the rivets were good. Bruce admitted to me he was too nervous to walk the beams, but knew he had a job to do. He sat down and straddled a beam and pulled himself across it to inspect the rivets at the far column. The rivets were all fine, but he marked two of them to be replaced, pulled himself back on the beam while still in a sitting position, and came down the ladders to the ground. He then told the construction superintendent, who had been watching Bruce the whole time, to replace those two rivets, and when that was completed, he would go back up and check the riveting. The super laughed, knowing that the marked rivets were perfect, but that Bruce was really saying all future ones had better be good also because, no matter how nervous he was, he would be checking every one of them. That story told me to make sure people were aware I would be checking everything I had to check, so they knew to do it correctly the first time to reduce the number of do-overs.[4-3]

As in many firms, partners ran little fiefdoms and had tremendous power compared with managers who weren't owners. The word was if one partner fired you, you could run to another to see if they would rehire you. Soon after I started, a partner turned 65 and had to sell his ownership interest to existing and newly created partners, as required by the firm's Partnership Agreement. I liked that dynamic whereby upper management consisted of senior partners, who would be cautious as they approached retirement, and junior partners who would be more aggressive. That balance had been missing from Brill Engineering Corporation.

CHAPTER 5

PROJECT ADMINISTRATOR

While working on projects, I frequently received directions or information from someone from the Staff Engineer Department. I wasn't sure what a Staff Engineer actually did, but they seemed important. One day, Duttenhoeffer and the head of the Staff Engineer Department, Art Jenny, called me into a meeting. Jenny said Staff Engineers managed projects and sometimes were referred to as Project Managers, and that one day I could become a Project Manager if I joined the department. It sounded interesting, and I agreed to the shift, even though I still didn't fully understand what a Project Manager was. Looking back, I think the idea of trying something new and challenging overcame my normal aversion to risk and sudden change.

Several Squad Leaders, more senior than I, were passed over when I was offered the opportunity to join the Staff Engineer Department. Once again, Duttenhoeffer had never said anything to me that he was considering me for better things, so I never expected anything. There was no such thing as a performance evaluation in those days at the firm; no one asked what you wanted to be in five years or what your career goals were. Indeed, no one even told you that you were doing well. The closest thing to a performance evaluation happened every other week when you received a paycheck. There also were no training programs as all development was by on-the-job training.

In the 1960s, the firm essentially operated in a matrix format[6]. Staff Engineers managed projects using personnel assigned by the four technical departments back then, namely the Highway, Bridge. Mechanical-Electrical, and Construction Inspection Departments. Project

---

[6] In lieu of a pyramidal hierarchy, a grid-like, matrix organizational structure is where staff have a dual reporting relationship with both a technical manager and a Project Manager.

personnel continued to be employees of their technical department, and it was their department head who determined when they would get promotions and salary increases, not the Staff Engineer/Project Manager.

Many Project Managers want to fill both project and general personnel roles because it gives them full control over the project personnel. I, on the other hand, was very comfortable with a matrix arrangement, with personnel assigned to others while working on my project so I could focus my efforts on the project and client. As Project Manager in a matrix organization, I didn't have to worry about such administrative minutia as why the size of anyone's office is smaller than someone else's or who covers when the receptionist is on leave. What I did worry about was our ability to stay within budget while we make the next deliverable on time with a product of appropriate quality.

A frequent problem in a matrix organization occurs when a department head unilaterally pulls personnel off a project and no one tells the Project Manager, who mistakenly assumes everything is going well. After getting bitten once, I learned I constantly had to check to confirm personnel were hard at work on my project. Also, that asking wasn't good enough as people often were too embarrassed to be truthful and admit they were working on someone else's project. The best way to find out what really was happening was to ask periodically to see what people had completed to date, such as working drawings or a report draft.[5-1]

In 1965, the Staff Engineer Department was organized roughly as follows: A very experienced Project Manager (Jenny) supervised one or two less experienced Project Managers, one or two relatively new Deputy or Assistant Project Managers, and an Assistant to the Project Manager (yours truly). Although, my title was Assistant to the Project Manager, I functioned as a Project Administrator. I was the youngest member of the Staff Engineer Department by 15 years. As the most junior, I was assigned project administrative tasks such as billings and collections, document control, budgeting, scheduling, and preparation of meeting minutes, subcontracts, and change orders. It's difficult to appreciate the importance of those tasks unless you've done them yourself. I strongly urge all who

24 THE ENGINEERING IS EASY

want to be a Project Manager to do administrative tasks at least once so they can review the efforts of those who actually perform them.[5-2]

CHAPTER 6

PROJECT MANAGER

After a year shadowing Jenny, I was promoted to Assistant Project Manager and then became Deputy Project Manager the next year. Another year later, I was named Project Manager on my first project which was for improvements to several blocks of Madison Avenue in downtown Albany, New York. I still was young and green compared with other Project Managers at Parsons Brinckerhoff. Fearing I couldn't make many mistakes or Jenny might replace me, I spent extra time and effort checking my own work. I developed lists of actions to take to minimize the likelihood of errors. Of course, I wasn't always successful and made my share of mistakes, but not so many or significant I was demoted.

The Madison Avenue project was a simple one and had an initial fee of only $500, but that didn't matter to me as I felt I had arrived. By the time that project ended a few years later, the fee had risen to $50,000, our margin (revenues less costs) had increased proportionally, and I believed I was a success. However, my euphoria diminished quickly when the client rejected our final invoice requesting release of retainage[7]. It seemed I'd never submitted a schedule for performing the work at the project's start as required by the contract. I had to create a now outdated initial schedule and formally submit it so the client would have one for its files, even though it no longer had relevance because the project had ended. Failure to submit that schedule at the project's commencement meant Parsons Brinckerhoff lost two months of cash flow on the $5000 in retainage (10% of the fee of $50,000). Back then, $5000 was a somewhat meaningful amount towards making payroll. In this instance, I'd missed something that should have been relatively minor, yet had serious consequences; next time I might not be able to resolve a mistake so easily. My error showed me the importance of creating a list of deliverables from the contract and checking off everything on the list as each was completed.[6-1]

---

[7] Retainage is money the client withholds until satisfied the firm has completed all obligations.

The most important lesson I learned from that incident was to read the contract, the whole contract, including all the boilerplate[8], and not just the technical scope of services. From then on, I read the contracts on my active projects at least once a month from cover to cover, including every attachment, appendix and reference document. My goal was to memorize the contract because I now realized a Project Manager must know its terms better than anyone else. It's a clear example of where knowledge is power. I also got into the habit each month of reading the entire chronological correspondence file going back to Day 1. It's remarkable all the things I'd forgotten until reminded by going through old files, such as commitments I and others made, but had not yet completed.[(6-2)]

Another lesson I learned from my first projects is engineers often underestimate the impact of property acquisition and utility relocations on a project's cost and schedule.[(6-3)] As an example, it was only when a highway contractor was preparing to construct a temporary relocation of a drain line, that we discovered no one had acquired the temporary easement required for its construction. It cost more time and money to get that easement than it would have if the need for the temporary easement had been realized prior to the start of construction. No one had overlaid the temporary construction plans with the acquisition plans to see if all the required property was scheduled to be acquired. After that incident, I made such overlays a standard requirement on my projects. Also, I made sure to add contingencies to both schedule and budget commensurate with the amount of property to be acquired and utilities to be relocated.

A somewhat similar incident arose on an initial site tour with the winning contractor on a project for the relocation of a half-mile of Green Street in downtown Albany. As the client, the contractor and I walked Green Street with the construction plans in hand, the contractor asked who is supposed to demolish that building on the centerline of the proposed construction. My heart sank as I looked at the one-story building in front of me that was not shown on the plans. Embarrassed, and feeling responsible for the

---

[8] Boilerplate refers to the various, administrative contract sections (invoicing procedures, liability limits, insurance requirements, affirmative action guidelines, etc.), often called the "small print."

plans' defect, all I could say was that we'll prepare a change order to add the demolition to your contract. No doubt, the contractor's charge for the change order was more than if the demolition had been in the original bid documents. The incident reminded me of my experience reviewing drainage designs when I worked for Rubin. It showed me that as Project Manager I must confirm the designers had toured the site with pre-final bid documents to be sure nothing was missed and not assume they had. It would be wrong if I had blamed the designers, because the error was mine for failing to check if they followed proper procedures.

Continuing education was not formally discussed in those days. Whenever I wanted to take a self-improvement course, I paid for it myself. I took a night course on systems analysis at the New School (a private research college in NYC) taught by staff from the RAND Corporation. The course helped me realize how often we look to solve a problem without understanding the basic issues behind the problem. Engineers tend to be linear, but we regularly encounter complex issues that require a consideration of non-engineering elements, including the social sciences. Until we appreciate all underlying elements, any answer may be ineffectual because it may be responding to the wrong base question.[6-4] Another thing I learned from the course was that when you're behind schedule and realize you need more resources, it's better to apply too much additional help than risking missing a due date with more, but still insufficient resources.[6-5] Stack the deck in favor of success rather than giving failure an equal chance to be the outcome.

In 1968, the American Society of Civil Engineers ("ASCE") was holding its annual conference at the historic Hotel del Coronado in San Diego. The program looked interesting and the chance to visit California for the first time was compelling. I went to Rush Ziegenfelder (Jenny's supervisor and a partner) and offered to split the cost with Parsons Brinckerhoff. Namely, if the firm paid my labor cost and the registration fee, I'd pay for the flight and hotel. Ziegenfelder agreed, provided I prepared a summary on the sessions I attended. I accepted his offer, and everything worked out perfectly. I still recall one item I included in the report, namely the Chief Bridge Engineer for the California Department of Transportation ("Caltrans") saying a bridge is inherently beautiful, and there's no need to gussy it up to make it prettier. Firms send many people to external conferences and seminars and should be diligent in requiring those sent to share new ideas and concepts they learned by preparing a written summary or holding an internal presentation.[6-6]

In those days, the firm consisted of multiple companies. In New York, it operated as a partnership (Parsons, Brinckerhoff, Quade & Douglas) because only individuals, partnerships or grandfathered corporations could be licensed to do professional engineering. Thus, all New York contracts were in the name of the partnership, although Parsons, Brinckerhoff, Quade & Douglas, Inc., which was not a grandfathered corporation and could not practice engineering, furnished all employees to the partnership. It was drilled into me by Jenny that we must write and say the appropriate firm name or risk increasing liability. After all, if we mistakenly wrote the corporation name on a letter transmitting design plans, the corporation may get charged with practicing engineering without a license. Also, our insurance wouldn't cover us in a law suit if the wrong firm was shown as the designer.

A few years later, the firm acquired a grandfathered corporation that could perform engineering services in New York and renamed it Parsons, Brinckerhoff, Quade & Douglas, Inc.; the partnership then ceased to take on new contracts in New York. However, it took several more years for the partners to begin acting as officers of a corporation, rather than owners of a small, privately held firm.

Jenny was almost fanatical that one should never make the same mistake twice and instituted rigid controls to check and double check quality. Whenever a mistake was made, a new control was added to make sure that mistake would never happen again.[6-7] Keeping to a commitment also was important to Jenny. When you said you were going to deliver something by a certain date, you had better do your best to deliver it.[6-8] If, however, you couldn't meet a due date, you were to warn your client that a deliverable date is going to be missed so the client could plan for the delay, say by rescheduling its internal review team. When you're going to be three days late, you told the client at least three days before the due date.

PROJECT MANAGER

Also, Jenny felt being punctual showed you respect the person with whom you're meeting.[6-9] Thus, when he scheduled a 7:30 am meeting in Albany, New York, it meant I left my Queens, New York apartment at 3:00 am to arrive at his house in Ridgewood, New Jersey by 4:00 am so we could reach Albany with time to spare. Moreover, if there was time before our first scheduled meeting, Jenny knew which of the client's employees started work early, and we would pay courtesy visits with them. In those days, government buildings didn't have security checkpoints, and you could enter through back stairways and wander halls to meet whomever you wanted to.

Jenny worked hard as well, with a strong work ethic. He often called and woke me before 7:00 am on Sunday to ask me for information. He'd been up for hours doing office work at home, and it didn't occur to him that I wasn't also. I guess he forgot I was single in my twenties, and it was likely I had gotten to bed only a few hours before.

Jenny was a stickler for form. For example, he made sure every letter was checked for errors before going out.[6-10] In those pre-word processor, pre-Xerox days, secretaries typed an original with several carbon copies and naturally were annoyed when a letter was returned for complete retyping because of one minor erasure or typo. Jenny taught me to understand that a sloppy letter or one with grammatical errors reflected badly on the firm and the letter writer. Moreover, typographic errors in a document imply its findings are suspect. One typo I caught was where a secretary changed the name of New York's scenic Seaway Trail to Sewage Trail. Another secretary described an attendee from the client's Environmental Department as being from the Mental Department. Now, I didn't catch every typo; I missed the one where the Power Authority's Chairman was described as Chairman of the Port Authority and had to scramble to recall a dozen copies of a white paper. It was very embarrassing.

I had to ignore my opposition to using White-Out correction fluid on documents during the 1966 NYC transit strike that lasted two weeks. During the strike, many employees had no way to get to work. We were finalizing a major deliverable to the NYSDOT in Albany, but the strike meant we didn't have enough secretaries in the New York City office to finish the paperwork. I moved to Albany for a week so I could use the three typists in our office there. The typists were too inexperienced to take the pressure of error-free work, and I had to permit the use of correction fluid on the typed documents or else we never would have finished the deliverable in time. Of course, I made sure the content was correct. By our standards, the documents looked somewhat shabby, but NYSDOT was understanding as they had a tight internal due date they had to meet.

I'm fairly competent in English grammar and it bothers me if a letter I receive has grammatical errors. Likewise, I worry the letters I send would be read by someone who knew the grammatical rules and would think less of me if they found a mistake. Thus, for example, I try to make sure pronouns have clear antecedents and plural nouns have plural verbs. Also, I probably spent more time than I should have correcting those drafting letters and reports who confused its and it's, effect and affect, due to and because of, or alternate and alternative. Even though I knew those mistakes really weren't significant, something in me (Jenny's teaching?) demanded I correct them.

Another thing Jenny taught me was that each letter should have only one topic.[6-11] Otherwise, the recipient might reply only to the easier topic and forget to recall there still is an open topic. In addition, the text should be clear what action we want the recipient to take (e.g., approve, respond with comments, or send us some information). Also, I learned from Jenny that you lack proof of all you accomplished if the letter transmitting a deliverable doesn't list every document being submitted and their unique title and document number.[6-12] Furthermore, every piece of project correspondence was to be uniquely labeled, placed in a chronological file and, then filed in multiple places so it could be found quickly.

Jenny frequently used the aphorism, "The Engineering is Easy," to remind us the hard part of most projects was not the engineering, but resolving the managerial issues and keeping administrative tasks in good order. For example, while answers to most engineering problems are relatively clear and definitive, solutions to problems involving co-workers, clients or project stakeholders rarely are simple. Most engineers are bright, causing some of us to coast along with our assignments. If Jenny had not trained me in his attention to detail, I would have breezed along doing everything way too casually and assuming my work was always error-free. Because of Jenny, I learned the successful ones are those who work diligently and proactively to avoid errors.

Complying with every one of Jenny's controls had two sides for it meant working in a very bureaucratic manner, albeit with fewer errors. Based on what I learned observing Jenny, my style of project management evolved to become very much into micro-managing. Since I knew all the pitfalls, who better than I should be involved to protect the firm's interests? Only I could write or call the client. No one could meet the client unless I was present. It was an approach that worked well for me for many years — that is until my first mega-project, Westway, which I'll discuss later in Chapter 10.

I created individual action lists for key staff working with me.[6-13] Whenever anyone telephoned me or came into my office about something, I'd pull out the action list with their name on it and ask about what they owed me. Too often, I found they had forgotten they owed me anything. Either that or they had put my request aside while working on something else without letting me know they would be late finishing the effort. My approach of continuously checking with them flows from "you get what you inspect, not what you expect."[6-14]

I collected articles from newspapers and technical journals I thought may be useful someday and filed them in such categories as speaking, writing, insurance, expert testimony, business development and quality.[6-15] Then, for example when I had to prepare a technical paper, I'd re-read the various articles on writing to get ideas of content and format to use. That way, I benefited from the good ideas of others and avoided common mistakes made by those writing their first technical papers.

Insurance was one topic I always felt deficient in understanding, including how it related to "liability" for our actions. When are we liable for something that went wrong, such as an error or omission ("E&O") in design or an accident to a motorist on a road we designed, and does our insurance cover us for what we did? To overcome my deficiencies, I talked with the firm's in-house attorneys whenever I sensed we were increasing our exposure to a lawsuit.

Similar to other consultant firms, Parsons Brinckerhoff was always being sued over one thing or another, and efforts defending our actions, while necessary, were a diversion of good people's time. The firm once was sued because a bridge it had designed in New Jersey "...did suddenly strike a car." Not that a driver at 2:00 am on a Sunday morning drove into the bridge, mind you. The suit had no merit, but we settled out of court for an amount less than we would have spent researching the facts.

Some clients look to shift liability exposure to their consultants as much as possible, even when unjustified. In 1984, Parsons Brinckerhoff was about to sign a highway design contract with a northeastern state Department of Transportation ("DOT") when the DOT added a clause that both increased the firm's exposure and required that we obtain insurance covering the increased liability. When we tried to buy insurance to cover the liability, we discovered no insurance carrier would sell such a policy. After describing the impossibility of buying the insurance, I requested the DOT delete the clause from the pending agreement. The DOT refused to make the change saying other firms are signing agreements with that clause and seemed not to care if those firms actually had purchased a policy. The DOT then threatened not only to cancel our firm's selection on the new agreement, but also to cancel all existing agreements we had with them if I didn't sign the agreement. Under great stress, as I was concerned the firm would lose millions in dollars of work and many employees would lose their jobs, I responded, "These other firms are fooling themselves and you, while we're being honest."

It took several months for the DOT to accept our position and delete the uninsurable clause from their agreements. In the meanwhile, our idle-time costs increased as staff were on standby waiting for the DOT to act rationally.

New York State's Office of General Services ("OGS") was more reasonable when there was a potential E&O claim in 1970. OGS wanted to build a 72" pipe which would pass under the location of a ramp we were designing for NYSDOT in Albany. We gave OGS the proposed ramp's horizontal and vertical geometry, and OGS set the pipe geometry. OGS then sent us the pipe geometry and asked us to confirm the pipe had sufficient headroom to pass under the future ramp. We reviewed the pipe geometry and said it was fine, and OGS constructed the pipe accordingly. A year later when construction of the ramp started, the contractor saw the pipe was higher than the proposed pavement and would be exposed. We re-reviewed everything and realized the pipe should have been designed to be three feet lower as it passed beneath where the ramp would be constructed. The error meant the pipe had to be removed and rebuilt to a lower elevation. OGS could have claimed against us for some of the costs to lower the pipe. However, OGS said while Parsons Brinckerhoff agreed with the pipe geometry, OGS's designer had our ramp geometry in hand yet had designed the pipe incorrectly. Because they had primary blame, OGS said they would cover all costs. That was very reasonable, but not likely to happen in today's litigious climate, which is why one can't be casual when reviewing the work of others.

Around 1970, Parsons Brinckerhoff lost the role as General Engineer for a toll road authority, a quasi-state agency. The story around the firm was the authority took the assignment from us when we refused to make a large donation to the governor's political party. According to the grapevine, the firm was told to inflate its fee to cover the donation so that it would cost the firm nothing, but we refused. It felt good knowing I worked for a firm that valued ethics and doing the right thing. That feeling was more than justified six months later when a Vice President of the firm that replaced Parsons Brinckerhoff as General Engineer pled guilty of making payoffs to win the assignment. Donating to a political party or a politician whose policies you agree with is acceptable, but paying someone as a quid pro quo for selecting your firm for work is wrong.

Marketing for work in those days generally was not very complicated and based on relationships between client and consultant. For example, NYSDOT's main office chose design firms to work on projects using a ranking of three firms prepared by their District Engineer where the construction was to be performed. One rural District Engineer, far from any large city, had a novel approach for recommending firms. When asked to list three firms for an upcoming assignment, he gave the first ranking to the firm that had visited him the most recently. That firm would stay first on all lists only falling to second ranking after another firm came to visit the District Engineer. A fall to third ranking and ultimately off the list happened as more and more firms came to visit. The District Engineer merely was rewarding firms that showed him respect by driving several hours to court him.

During the 1960s, Parsons Brinckerhoff worked on numerous projects for the City of Albany, New York. There were many engineering firms in Albany, but our firm often was assigned the complex projects, such as serving as General Engineer for the Albany Airport. The Mayor of Albany was Erastus Corning 2nd, who was the longest-serving mayor in the country. If you wandered into City Hall to admire the building and the Mayor's aide saw you, he would come over and invite you to meet Mayor Corning. You'd never think the Mayor, a tall patrician man, was a died-in-the-wool politician when you met him.

One day, I received a call from the Mayor's office asking to meet with him. Mayor Corning wanted to reassign twelve properties from one council district to another, and our task would be to revise the council maps. He became frustrated when I showed him the properties were not adjacent with the council district to which he wanted them moved, and there was no way to make it work without changing several other district limits. He wasn't satisfied; he rolled up his sleeves and tried to move district lines around to suit his plans, until he realized he couldn't get the result he wanted. The whole process showed me I should never forget politicians worry about such seemingly small things as to which council district twelve properties are assigned.

Although I was assigned to the NYC office (where all the designs were performed), I spent almost half my time in Albany for several years while working on various state and municipal projects. There were 120 inspectors and surveyors based in Albany at peak during that period. The head of the Albany office, call him Sam, was a quiet man in his sixties and a good, honest person. He never said much in meetings, but frequently nodded or frowned during discussions. He came across as a calm figure of sound judgment and a wise sage knowledgeable of heavy construction.

After I got to know him better, I realized he never had much to say because his knowledge in some areas wasn't as deep as one assumed just by looking at him. With that realization, I knew it was best to use him to take advantage of his strengths, while limiting his involvement in areas where he was weak. More importantly, I learned from Sam that it's a good idea to keep your mouth shut if you are in over your head.[6-16]

In 1967, Jenny asked if I would relocate to Albany if Parsons Brinckerhoff opened a design office there. That was the second time I had been asked if I would relocate. I agreed because it was likely that relocating, while somewhat of a hardship, would enable me to advance more quickly than staying in one place. However, the firm ultimately decided not to open the office, and the relocation never came to pass. The whole process of considering a relocation, agreeing to relocate and then waiting and wondering if the relocation would happen was nerve racking.

Office Relocation:

The NYC skyscraper at 165 Broadway, where I worked, was to be demolished in 1967 along with an adjacent skyscraper, the Singer Building (at one time the tallest building in the world). A single skyscraper would be constructed in their place to house US Steel offices. The demolition of our building meant we would relocate several blocks away to 99 John Street. It was hard for me to understand why someone would destroy two good buildings, with the concomitant loss of two years of tenants' rent, just to replace them with a similarly-sized building. However, it made sense when it was explained the new building would generate much more rent per square foot because it would be constructed with modern amenities, such as high-speed elevators, central air conditioning and enhanced electric service to accommodate large computers. No doubt the developer ran the numbers, and estimated they would recover their investment back in about seven years — a typical return period in real estate.

A key concern of our move to John Street involved the relocation of the computer system such that our work efforts wouldn't be impacted too much in the interval while the old system was shut down and before the new system was activated. The Computer Manager (call him Rudy) handled the relocation. Our computer system in those days was a large mainframe, as personal computers didn't exist yet. These large computers took up a whole room which had to be air conditioned, or the computer would overheat. If you needed a program run, you filled in a work order and give it to Rudy who would complete it in the order received. It often took a few hours for the simplest of calculations to be returned to you.

Rudy did a good job coordinating the move, and I was surprised to hear he wouldn't continue in the new space. The rumor was that someone discovered, just before the move, that Rudy occasionally was using the firm's computers after hours to sell computer services to other firms for his own enrichment. Had Rudy made an error in the work he did for another firm, it's likely Parsons Brinckerhoff would have been sued by that firm — a major reason why many firms prohibit employees from moonlighting. Once the relocation was complete, Rudy's supervisor told him we knew what he had done and allowed him to resign. There always will be those looking to game the situation, and I now realized the importance of making sure no one reporting to me fell into that category.

CHAPTER 7

MANAGING A MULTI-DISCIPLINARY PROJECT

In 1967, I was Jenny's Deputy Project Manager on the design of I-787, a new Interstate highway spur from the NYS Thruway to Albany, New York. It was among the largest projects NYSDOT had underway, and the work was divided into several, sizable, multidisciplinary, construction contracts. I replaced Jenny as Project Manager in 1971, when he began managing Westway. The complexity of the I-787 project would have been too much for me to have been the original Project Manager, but after six years as a Staff Engineer with Jenny, I was prepared to take it over midstream. Jenny's willingness to develop me to succeed him showed me the value of training a successor so the Project Manager could move on to their next project opportunity. Following are discussions of issues I faced while gaining experience managing this large, multi-disciplinary project.

The I-787 project included several disciplines in which I had limited strengths, including electrical engineering for light fixtures, mechanical engineering for pumps, port engineering for bulkheads, and architecture for modifications to a postal building. I had to learn enough about each discipline to be able to coordinate everyone's efforts and have credibility as Project Manager. While I didn't know how to design, say, street lighting, I reviewed every electrical plan to see if it made sense. For example, I checked that the dimensions agreed with the highway plans, the plan presentation met client standards, and there were no typos. More often than not, I'd find something had to be changed in every discipline's plans.

Numerous Delaware & Hudson ("D&H") Railway facilities had to be relocated, including a crane yard and several operating tracks, to make room for constructing I-787. Our scope for I-787 included designing the railroad facility relocations and incorporating their related construction in highway bid packages. For one long stretch, the D&H mainline would be relocated to operate in the median of I-787. We were concerned that locomotive headlights would temporarily blind and panic automobile drivers in the opposite direction and decided to conduct nighttime tests to assess the glare's significance and whether high screens would have to be constructed to protect against the glare. The temperature was minus 27°F with a brisk wind on the night the D&H made a locomotive available for tests in their Saratoga, New York rail yard. I wore the warmest, winter-weather clothing I had, including long johns, but nothing helped given the bitter cold. I doubt I ever felt colder as the tests dragged on for three hours. We made numerous measurements as the locomotive moved back and forth on its track, so we could do our analysis. The evening was successful in that the analysis enabled us to design the appropriate screen, and no one developed frostbite.

Back then, little of Parsons Brinckerhoff's long history in railroad design was recent, and I was concerned potential clients seeking a railroad engineering consultant would be unaware of the expertise we gained on the I-787 project. To demonstrate the firm's railroad experience, I created a brochure page entitled "Relocation of D&H Railway Facilities," highlighting Parsons Brinckerhoff's role in railroad engineering. The page described only the railroad engineering portion of the scope in the contract, without mentioning the balance of the contract scope, which was overwhelmingly highway engineering. Comingling the railroad effort in the I-787 highway experience brochure material would have increased the likelihood potential clients wouldn't appreciate the extent of the firm's current railroad engineering experience. I also added a separate paragraph on my résumé under the subheading "Railroads" so my personal railroading efforts could be recognized.

Because construction of I-787 would eliminate several railroad grade crossings of city streets, approval of the Public Service Commission ("PSC") was required to remove those grade crossings from the books. I was to provide expert testimony in Albany at the PSC hearing on the soundness of the design eliminating the crossings and drove to Albany in my Chevrolet convertible the night before. The next morning, I checked out of the Sheraton InnTowne Hotel after breakfast and went to get my car to drive to the hearing. No car! It had been stolen overnight. I called the police to file a report and then hurriedly took a taxi to the hearing where I testified almost in a daze. Fortunately, a hearing to eliminate, rather than add, a grade crossing is somewhat perfunctory. By putting the theft out-of-mind as best as I could, I made it through my testimony and ultimately everything was approved. I had to take the bus back to New York City. As for my car, it was recovered a week later, slightly banged-up, but drivable.

After that incident, I generally used rental cars for business travel so I wouldn't be stressed worrying about my personal car. There were a few company cars assigned to those who traveled a lot. To cut costs, the cars didn't have air conditioners, radios or clocks. Jenny taped a transistor radio to the dashboard of the car he was given to drive. I did get a company car a few years later with a/c, a radio and a clock, all of which had become standard issue by then.

Design of I-787 happened during the Vietnam War, and there were frequent military convoys using the Thruway to and from Camp Drum in upstate New York. Delays caused by encountering those long convoys often added an hour each-way to my drive to Albany, which meant the trip was very tiring. To stay fresh and alert, I often took a three-hour bus ride to Albany rather than driving. Whether by bus or car, I had to leave early to make sure I wasn't late for a meeting. There were few hotels and motels once you left big cities in those days. Sometimes, I had to share a motel room bed with Jenny given the paucity of places to stay in Albany.

Constructing I-787 required a permit from the Army Corps of Engineers ("COE"), because a portion of the highway was to be placed on newly constructed fill in the Hudson River to minimize property acquisition. I visited the COE in NYC to describe the proposed construction and learn what they would require to approve the permit application. The COE lieutenant I spoke with was very helpful, and eventually the project received a permit. However, the lieutenant indicated he was upset that the highway would encroach in the river. It was the first time I appreciated the extent to which highways have environmental impacts, a topic that was only beginning to be mentioned when discussing alternatives.

Parsons Brinckerhoff's scope for I-787 included construction inspection of several construction bid packages over a 15-year period. While we were designing the last design packages, the firm's Resident Engineer, Jim

Bruschi, inspecting the construction of an earlier package, called to complain about the quality of electrical design for roadway lighting and lighted signs. Rather than resolving the issue by telephone or letters, I decided to bring our Chief Electrical Engineer, Lou Benjamin, to Albany and have him learn about design concerns directly from Bruschi. A week later, I got them face-to-face in the field trailer for what I felt could be a contentious meeting.[9]

I imagined Bruschi pointing out numerous design shortcomings making construction difficult, while Benjamin defended his design. The meeting started, and Bruschi mentioned some design tweaks he'd like to see that would make his job easier; nothing was very significant. All in all, the meeting was short and sweet with both parties acting collegial and no hard feelings about anything. Benjamin came away with some minor items to consider in future designs, and Bruschi came away knowing he could call Benjamin whenever he had an electrical design issue he wanted resolved quickly. As for me, I learned office and field people think differently and getting them together periodically is worthwhile, if only for each to see that the other really wants the same thing — a successful project.[7-1]

Another time, Bruschi called to say the construction contractor was asking for prompt approval of a minor change to a bridge we designed. I asked John Kalapos, the senior Bridge Engineer for the project, to review the request. Kalapos came back the next day and said, "We took months to do that original design and must have had a good reason, but no one recalls why we designed it the way we did, after only one day's review. On that basis, we shouldn't agree to a change since we could be making a serious error without understanding all the facts."

I then told Bruschi, that unless the contractor could give us a few weeks for a proper review of the potential change, it should complete the construction as originally designed. As the contractor didn't want to wait, they built it as designed, and everything went well.[7-2]

---

[9] I took my first airplane trip to that meeting. It was on a WWII DC-3, still in use some 25 years after the war ended.

I remembered that incident some years later when I heard a consultant firm faced an E&O claim on a project when a vertical wall of a U-shaped, open drainage channel failed during a flood. Someone from the firm, not the original designer, had approved a request from the field during construction for a change in reinforcing steel where the wall joined the foundation without knowing the original design load on the wall was based on unusual flooding conditions. In the end, the firm and its insurance carrier paid several million dollars for remedial repairs because no one asked the original designer if the proposed change made sense.

Because it's unlikely the contractor and construction inspector fully understand every reason behind the designer's intents, field personnel may make a seemingly minor revision of a designed element not knowing the designer strongly intends for that element to remain as designed. To reduce the prospect of field changes overriding design intent, we created a series of notes for the field personnel for each I-787 design package.[7-3] The notes for the field described items that may appear flexible, but are not. For example, a note might state that a specific element in the current contract must be constructed exactly in the location shown on the plans to avoid conflict with a future project.

Mall Arterial:

New York Governor Nelson Rockefeller was said to be embarrassed when the future Queen of the Netherlands, Princess Beatrix, was driven through Albany's red light district on the way to the Governor's mansion. The Governor decided Albany, New York's capital, needed to be improved to eliminate the worst parts of the city (referred to as the "Gut"). The focal point of the improvement was the South Mall, a proposed complex of state buildings adjacent to the mansion. Constructing the South Mall required acquisition of many properties in the Gut, a form of urban renewal common in that era. Parsons Brinckerhoff was selected to design the Mall Arterial, the highway that would connect I-787 to the South Mall.

The Governor appointed retired General Cortlandt Van Rensselaer Schuyler to oversee development of the South Mall and the Mall Arterial, and Schuyler held monthly meetings of key participants which Jenny or I attended. At one meeting, there was a discussion about how to haul very large beams safely through Albany's streets to the project site. The somewhat comical dialogue between Schuyler and Police Captain Devlin, the City of Albany's traffic coordinator, went something like this:

> Schuyler (in his typically commanding voice): The next item on the agenda: Does the city have any special requirements for hauling very long, structural beams through the streets?
> Devlin: You'll need a police car with flashing lights leading each truck hauling a beam.
> S: Very good, and the state will reimburse the city for the driver and car. The next item on the agenda is ....
> D: Well, you'll need a second policeman in the car in case of emergencies.
> S: That's fine. And the next item ....
> D: And of course, I'll have to be on duty to coordinate everything from headquarters.
> S (somewhat weakly): Fine.
> D (sensing the General was on the ropes): And you'll need a second police car trailing each truck.
> S (more weakly): Okay.
> D: With two policemen in it.
> S (slightly sarcastically): Naturally.
> D (applying the final blow): And, of course, all police will be doing this assignment outside their normal duties and will be paid overtime rates.
> S (very weakly): Of course.

The previous exchange is a perfect example of what's called the salami tactic of negotiating: Once someone agrees to buy the salami, slice a little piece off it. Then, slice another piece, and another piece, etc.

The state's architect wanted to avoid placing overhead lights on the Mall Arterial, because he felt standard light poles would mar the view of drivers approaching the imposing buildings he was designing for the South Mall. To satisfy the architect, a NYSDOT lighting engineer directed us to incorporate a highway lighting system on the Mall Arterial he had seen

MANAGING A MULTI-DISCIPLINARY PROJECT   43

installed in California's Bay Area. In that system, fluorescent lights were placed in rail barriers along the roadway's shoulder. After reviewing the system constructed in California, our lighting engineers told NYSDOT that the operating cost for the system would be very high, and the use of the Bay Area ballast would result in low lumen output in the frigid temperatures of Albany's winters. However, NYSDOT said that they were fine with everything and directed us to design the lights in the rail barrier.

Since all the rail light tubes, housings, lens and ballasts would be custom-made, we developed specifications for testing the manufacture of the various lighting elements. My graduate course in Probability & Statistics helped me determine how many items to test for each element, and how many failures per element tested would be allowed, to satisfy ourselves everything was manufactured properly. The required number of tests and maximum number of failures were included in the bid documents.

The low-bid contractor selected a manufacturer in Oakland, California to produce the rail lighting elements. A senior Electrical Engineer from our San Francisco office joined me in a visit to the manufacturer, and we oversaw the testing of the various elements to see if they complied with the specifications. However, while the specifications I prepared stated the lens had to be able to survive an accident in cold weather, I had failed to describe a test to prove the lens was acceptable. With the manufacturer's concurrence, we invented a test as we tried to replicate what might happen if an automobile sideswiped a lamp lens in frigid weather. We put the lens in a tub of ice for several hours and then struck it firmly several times with a hammer to see if it would shatter. It survived without a crack, and we deemed it satisfied the intent of the specifications.

There was a problem with a proposed street sign ("No Parking Here to Corner") that was about one-foot square to be installed in Albany as part of the I-787 project. The design plans had the sign attached to a new 8-inch wide-flange beam to be set in the sidewalk. After it was constructed, the post looked ridiculously overdesigned supporting such a small sign. When I asked our Structural Engineer why he selected the post he did, he said that's what the computer program showed was necessary. The fact that probably every other similar sign in the country was placed on a 3-

inch channel meant nothing to him, as he defended his design. He didn't consider a large, inflexible sign-post anchored in the sidewalk is a serious hazard should a vehicle jump the curb. Rather, he thought spending 20 times the replacement cost of an inexpensive channel to build a sturdy post made sense. It's common practice to overdesign an item whose failure results in significant damage. In this case, overdesign resulted in a more dangerous end product and was totally unwarranted. The lesson here is not to blindly accept a computer's output, but to look at the result to see if it passes a sanity check.[7-4]

Snow Removal

Snow removal by snow plows and dump trucks would be difficult on the five-level interchange ramps between I-787 and the Mall Arterial, as there was no room to push snow to the sides of the ramps and snow dumped over the bridge parapets would fall on drivers below. Dump trucks would have to travel long, circuitous routes to haul snow from the plows on the ramps, some of which are over 100 feet in the air, to a dump site before returning to the plows for the next load. The client asked me to study snow removal methods other than the use of plows and trucks and to compare the cost and effectiveness of all options.

It was the first formal report I prepared, and it was on a rather esoteric topic for a Highway Design Engineer. I researched the literature, talked with many people, and identified several options including radiant heating in the bridge parapets and both electrical heating wire and hot water pipes in the pavement. I ultimately recommended staying with plows and dump trucks because both the initial and life-cycle costs of the high-tech options were very expensive and their snow removal effectiveness dubious given the technology that existed then. Also, I knew that when snowfall was heavy, it was likely the ramps would be closed and traffic diverted to street level. That meant there was no reason to rush snow removal on the ramps, and using dump trucks wouldn't inconvenience drivers. The client accepted the report, and I felt my research had advanced the state-of-the-art. In retrospect, had I really been on-the-ball, I would have published a paper on the topic, so others could benefit from my research and I would receive acknowledgement for my efforts.

Pre-Excel:

I created a computer spreadsheet years before programs such as Microsoft Excel software had been created, but wasn't perceptive enough to realize what I had and to copyright or patent it. In those pre-Excel days, the NYSDOT process for preparing engineering estimates for a construction bid package was very convoluted. The estimate started with identifying all construction items and their unit costs, and then estimating the quantity of each item.

Hand calculations began once items were identified, and their quantities calculated by funding source. One I-787 construction package required about 10,000 hand calculations, with everyone a potential for a mistake. Next, each of the 10,000 calculations was transposed (another potential error) to a series of sheets with funding source in the columns and items in the rows. Today, it would be a series of inter-connected spreadsheets. It took several engineers a week to do the calculations for I-787 before the total of the rows equaled the total of the columns. As you can imagine, there were times when the row total vs. the column total was off by only a few dollars, and everyone would be scrambling trying to find and correct that last error. Once all totals were in agreement, the 10,000 numbers were typed onto pre-printed forms-still another potential error.

It was mid-day on a Saturday afternoon, and the row vs. column numbers for the I-787 package were only $80 off, when I went home thinking the team I left behind should finish the estimate in an hour or so. However, I was wrong, as they found even more errors, rather than resolving the $80 discrepancy. Because we had to complete the estimate that weekend, four employees came back on Sunday, which was Mother's Day, to finish the estimate. I was very upset they had to work on Mother's Day and resolved to find another approach for the future.

I was sure using a computer was the answer and started working with a programmer. It took a few months, but we finally had a spreadsheet program that worked. It worked so well, in fact, that I used it for other purposes, including overhead, project and proposal budgeting. The program had a broad versatility, but I couldn't get others to appreciate its value. Now this was back in punch card days, and when the firm bought a new computer and converted away from punch cards, the Computer Department Manager unilaterally decided this program was not worthy of reprogramming. Thus, one day when I went to use the program, I was taken aback to find it no longer existed in our portfolio of internal programs. Frustrated, I had to complete my effort using paper, pencil and calculators. And, within a few years, the firm was buying spreadsheet programs such as CalComp and Excel.

CHAPTER 8

MANAGING MULTIPLE MULTI-DISCIPLINARY PROJECTS

By the mid-1970s, I managed more than just highway projects, having been designated Project Manager on bridge, traffic, water supply, sanitary facilities, power, environmental, and planning projects. While my technical strength in some of those disciplines was at best average, the skill required to manage the project elements was relatively similar, as the basic principles of scope, schedule and budget applied irrespective of the primary discipline. In other words, having gained experience and understanding of cross-disciplinary strategies, I was ready to manage projects of any discipline. Even as the portfolio of projects I personally managed increased in number and size, I often served concurrently on other projects as a deputy or in a key role.

Whenever I started a project, I'd request a budget from the internal departments and external subconsultants involved with the project. There are always groups who ask for more budget than they really need, because they want to beat the budget and look good. Meanwhile, others give you a budget they think will make you happy and then ignore the agreed-upon budget, often exceeding it by the time they finish. Unfortunately, more people exceed their budgets than beat them. Until I'd worked with someone for a while, I didn't know if they're the ones who will usually beat their budget or usually exceed it. In any case, I always got budgets from anyone who would work on the project as even approximate budgets can be used to assess progress.

After receiving budgets, I customarily reduced them by 10% and kept the excess as a contingency in my back pocket.[8-1] Some groups would complain they need every penny, but I generally settled them down by explaining that, "While some groups find they don't need their entire budget, others need a little more. For example, one group could finish on budget, but later have to rework some efforts because of a change that ripples through the project. If I control the contingency, I can give extra money to whichever group really needs it. Also, if you overrun a budget by a little, I won't say anything, so let's set the budget a little low as a stretch goal."

Internal meetings of key project staff proved invaluable not just for sharing information, but more important for reducing surprises.[8-2] Once a project was active, I'd schedule a meeting every week and invariably learn something I hadn't anticipated. Interfaces between disciplines or offices increase the risk of something going wrong, and these meetings gave me the chance to surface and address a minor problem before it became a critical one. At each meeting, I'd ask everyone what they're doing and what they needed to finish their efforts. Often, I would discover one party failed to tell another they've changed something or someone never mentioned they're expecting to receive something by a certain date. I'd also hold a monthly all-staff meeting so everyone had a chance to understand what's happening and how their efforts fit into the overall picture. The more even the most junior staff know, the more they feel invested in the bigger picture and can boast to family and friends about this wonderful project on which they're working[10].

In 1971, NYSDOT had a financial crisis and terminated many consultant contracts, including several of Parsons Brinckerhoff's, two of which were sizeable I-787 construction inspection agreements. The contractors continued constructing those portions of I-787, with NYSDOT now performing the inspection. Altogether, we had to lay-off 50 design staff in NYC and 100 inspection staff in Albany.

Prior to the agreement terminations, I treated interim billings on a fixed price (lump sum) contract casually, assuming it would all work out in the end. I never worried if our in-progress billings were slightly ahead or behind the earned fee because I knew when we completed the last of the scope, we would collect the total fee in the agreement. With in-progress billings, we approximated our percentage complete and were paid that

---

[10] The old story goes like this: Separately, I approached three construction workers doing the same task and asked what they were doing. The first said, "Laying stones," the second said, "Building a wall," and the third said, "Building a cathedral." Now, which one do you guess is the most motivated?

percentage of our fee. However, termination meant we no longer could complete our scope, and what was invoiced and collected on a provisional basis had to be revised to reflect the actual value of work performed. Any inability to prove what we had accomplished would reduce our compensation, and now, we had to prove we completed (say) 40% of the work, to be paid 40% of the fee. I read the termination clauses in the cancelled agreements more carefully than ever before. Fortunately for the I-787 project, I had used a document control system that quickly enabled me to substantiate everything we had done, including the number of plan sheets and reports prepared and meetings attended. That made it easy to prove to NYSDOT that the final earned fee we were entitled to was, by and large, the same as the amount we had provisionally invoiced. Following Jenny's attention to detail paid off again.

Even though Parsons Brinckerhoff's construction inspection contracts were terminated, manufacturing of rail lighting elements for the Mall Arterial and I-787 was continuing in Oakland, California. NYSDOT now realized our terminated contracts provided the mechanism for our firm to determine if the forthcoming manufactured elements were in accordance with the specifications. NYSDOT asked if we would resume testing under a new agreement. We said that based on the way they calculated compensation, our reimbursement for testing $1 million worth of equipment would be only a few thousand dollars and our profit would be a few hundred dollars. Given the liability risk we would assume for the testing was in the $1 million range, we asked for a limitation of liability[11] commensurate with our reimbursement and profit. Also, we explained that while we normally would provide a service even at a loss to a good client like them, we hoped they would understand the risk was just too much in this instance. The risk would not have been an issue if our terminated contact were still in force as there was more than enough other work for us to assume the risk on this small effort. NYSDOT said they couldn't limit our liability, and in the end, did the testing themselves. Note, a few years after construction was completed, NYSDOT determined the energy

---

[11] Limitation of liability sets the maximum amount the firm would owe NYSDOT as compensation for an error or omission.

costs to operate the rail lights were too high (as we had warned them during the design phase), and the lights were permanently turned off.

The fiscal crisis was devastating to Parsons Brinckerhoff, as NYSDOT was a major client. Not only was the firm forced to lay-off 150 employees, but the fiscal crisis meant there would be fewer new DOT projects for some time. All our New York competitors faced the same problem, and now many firms were chasing the rare DOT project that came along. By far, the largest potential DOT project in 1971 was the Westside Highway ("Westway") Engineering Management project in New York City with consultant engineering fees exceeding $100 million. Westway would be a new Interstate highway in Manhattan alongside the Hudson River. All major engineering firms pursued that project in a feeding frenzy, but NYSDOT ultimately selected Parsons Brinckerhoff for the work. Within a few months, several of the firm's officers and key senior managers, including Jenny, who became the Westway Project Manager, and me, went from partially to fully billable. Fully billable means every hour of an employee's labor is paid for by a client, and there's no idle time assigned to overhead.

When Jenny started managing Westway, I became the project's Manager of Project Controls, in addition to my serving as Project Manager on the still active agreements for the I-787 project. As Westway's Manager of Project Controls (in essence, a senior-level Project Administrator), I oversaw contracts, subcontracts, invoices, document control, cost and schedules. Project Managers on large projects generally focus on keeping their client happy and the various project technical leads on track, leaving them little time to do project control tasks properly the way a Manager of Project Controls can. The Manager of Project Controls often is the exception reporter, advising the Project Manager on areas of concern (e.g., we're trending in the wrong direction on costs, or the following deliverables are due next month). Over the years, I frequently designated the Manager of Project Controls as Deputy Project Manager, having a good appreciation for the value of that position.

Until the early 1970s, most American public sector clients selected a consultant firm sole-source (i.e., without competition) for assignments based on what the client knew of the firm. By that approach, clients never asked who would be designated Project Manager as clients knew what the firm could do and that was all that mattered. However, the selection process for federally-funded projects changed after a 1973 scandal in Maryland. That scandal resulted in Spiro Agnew resigning as U.S. Vice President, because he failed to pay taxes on monies related to engineering firms paying bribes for work when he was Maryland's Governor.

To reduce the potential for bribery, new federal regulations relative to consultant selection were created after Agnew's resignation. These regulations required that almost every federally-financed project have a competitive selection process, typically involving qualification packages and presentations/interviews. (Note most major state transportation projects were at least partially financed with federal funds.) This sea change in the consultant selection process meant clients began asking competing firms for their proposed Project Manager's name and résumé.

Unfortunately, many public sector clients didn't fully understand a consultant's Project Manager's responsibilities and were unwilling to take a chance on a Project Manager unskilled in the project's primary discipline. This attitude meant it was more difficult for someone like me to be designated as a Project Manager on non-highway projects. Many consulting firms now were forced to designate engineers, experienced in the primary discipline, but without project management skills, as Project Manager. It then became necessary to back up these somewhat weak Project Managers with someone capable of dealing with the various management and administrative issues. Because many clients questioned the need for such additional labor costs, it was a lose-lose proposition when a consulting firm couldn't convince their client of the necessity for the additional support. Now that most Project Managers would come from the design departments, Parsons Brinckerhoff upper management eliminated the Staff Engineer Department concept.

As mentioned, I was Westway's Manager of Project Controls. A year after we began that project, the project control systems were functioning well, and I left Westway when a Project Manager position on a New Jersey toll road project in Middlesex County opened up. The negotiations for design of the toll road project were very protracted before the client and we agreed that the design fee would be based on a percentage of estimated construction cost. I should have appreciated that protracted negotiations prophesize that future claim resolutions will be unfruitful — which they were. Immediately after we signed the contract, the owner made several changes that increased the construction cost by 25%. It was evident they knew of these changes during negotiations, but sat on them until we agreed to the design fee. When we asked for a fee increase commensurate with the construction cost increase, the client said you're not entitled to relief under the contract terms. Without a fee increase, we lost money completing the project

It was my first losing project, and I agonized afterwards over why we couldn't have done better. My first reaction was that I learned one client to be wary of in the future, but that excuse blames others. It's better to learn what I could have done differently. Upon reflection, I realized the contract should have included more specifics about the project we thought we were negotiating to design. Had I insisted on defining details of the smaller, original project in the contract, it's more likely our request for a fee increase would have been successful.

While on the topic of negotiations, let me recall a client's Project Manager, call her Fran, and my experience with her. I was preparing to negotiate the first change order on an existing contract, when Fran called me in for a meeting to complain about missteps on our part and implied she might terminate the contract if our quality didn't improve. As I knew there had been some things we could have done better (there always are), the need-to-improve discussion meant we were handicapped in the subsequent negotiations. Several months later, just prior to our next negotiation for a change order, Fran berated us for another miscue. And as we prepared for a third negotiation the next year, I said to my deputy that I bet it would happen again — and it did. I had discerned Fran's negotiating tactic was to make you feel you were lucky not to be thrown off the project.

It's important to sense early-on in a business relationship, if the other party is an advocate of a negotiating tactic such as salami or you're-lucky-I-don't-terminate-you. No such approach is ethical when one is trying to

MANAGING MULTIPLE MULTI-DISCIPLINARY PROJECTS

develop a long-term relationship of equals. However, failure to identify the other party is using such an approach leaves you at a disadvantage.[8-3]

The New Jersey toll road project was the first one where I reported to Bill Dyckman, a partner of the firm. Jenny and Dyckman had very different styles. Jenny was always multi-tasking, trying to complete every item on his lengthy to-do list, today, if possible. Dyckman was more measured, focusing on his highest priority until satisfied it was done properly. Dyckman called his style the Theory of Limited Objectives. I realized both approaches had merit in different situations and moved back-and-forth between the two over the years.

Dyckman was highly respected by competitors, who often sought his advice on issues affecting the profession. He had a slow, deliberate way of speaking as he searched for the perfect word or phrase to convey his meaning. His cerebral style frustrated those who were impatient, and I frequently saw individuals jump into one of Dyckman's long pauses to complete his sentence. They would do that only once, as he merely would give interrupters a brief look of almost disdain and start whatever he was saying over from the beginning. Dyckman felt his thoughts were valuable, and he wasn't going to let anyone else finish them for him.

I was unsophisticated back then about the relationship between project management and general management of the firm. For instance, interest rates in that time period were running well over ten percent, while the firm's average profit on projects traditionally had been about four percent. I asked Dyckman if the firm wouldn't do better investing in CDs paying 10% than doing projects. Dyckman replied the firm was earning more like 15 to 20 percent when clients paid on time. He explained that revenues (costs plus profit of four percent) billed and collected in three months after expenditure equals 4% x 12/3 or 16 percent profit per annum. The reason why upper management was always concerned about timely billings and collections now made perfect sense to me.

Ten years later, Dyckman announced he was retiring from the firm. Before he left, I asked him for advice about an issue I was having with someone else in the firm. I knew he always had something to say about this topic, and it was very discouraging, when Dyckman offered no suggestions for me; in fact, he seemed disinterested in the whole matter. It was only when I approached my own retirement phase that I realized it's natural that those who are retiring will care less about issues that once were important to them as they begin the process of letting go.

Parsons Brinckerhoff was pursuing a major light-rail transit program in Denver in 1975, and I was asked to relocate to Denver to be Manager of Project Controls if we won. I wanted this assignment because I hadn't worked on any transit projects and felt it would broaden my value to have one in my portfolio. I knew the Project Manager and his three deputies we were proposing for the project. Each, while technically capable and personally pleasant, was somewhat laid-back, and I was concerned that none of them were hard-driving enough for the program to succeed — that control of quality, budget and schedule might be lax. When I raised my concern with the designated Project Manager as to who would be the enforcer, he asked if I would fill that role. I said it was okay with me; I just wanted to make sure someone had that responsibility. In fact, I feel every project needs at least one person who asks the tough, internal questions of the project team.

My wife Mary was looking for a career change at that time and agreed to become the project office Administrative Manager if we won the Denver assignment. Relocations can be hard on a family, especially when you have a two-year old. My parents were getting ready to retire and said they would move from Queens, NY to Denver to be with us, which would solve our babysitting needs. Now, we put our lives on hold as we waited to see if the firm was successful in its pursuit.

A few months later, we learned the firm was selected for the assignment, and I told Mary and my parents we would be moving to Denver in three months. However, the three months stretched out as the program languished while the client tried to get the federal financing they were counting on. More waiting for everyone — Mary still not looking for a new job, and my parents wondering where they would spend their retirement years. Six months later, I still was getting tired marking time waiting for the Denver project to start when a former client, the Power

Authority of the State of New York ("PASNY"), requested I manage a new project for them. At that point, our designated Denver Project Manager released me from my commitment to work on his project, and I was able to take over the PASNY project. Working in New York turned out to be a good outcome for my family and me, as the Denver project folded several months later when the client became resigned to the fact that federal funding was never coming through.

Power Authority of the State of New York:

PASNY was planning to construct a nuclear power plant in rural Greene County, New York. To benefit the adjacent hamlet of Cementon, PASNY offered to build the hamlet a water supply system and a sanitary treatment system. PASNY retained Parsons Brinckerhoff to plan and design the works, and I was the designated Project Manager. We completed the initial planning phase that showed Cementon had no need for sophisticated water and sanitary systems, as local wells and septic tanks were all they would require for decades. At that point, we had expended 25% of the agreement fee. I stopped our efforts and recommended that PASNY terminate our assignment, as there was no reason for us to continue working on the two systems. Parsons Brinckerhoff's upper management supported my decision not to waste a client's money, as proof of the firm's commitment to do the right thing. PASNY was disappointed they couldn't help Cementon, but pleased we stopped of our own volition and accepted my recommendation to cease work.[8-4] Thereafter, whenever someone said all consultants are money grubbing (and it happened often), I'd mention the decision on that project to stop work and save the client money.

Failed projects, such as the Cementon water and sewer projects, raise the question: Is there a due diligence test a Project Manager should perform on Day 1 to be able to recognize a project's likely to hit a wall requiring work to stop? My answer is, at the start, the Project Manager should identify potential risks leading to project abandonment and then develop strategies to address each of those situations. Strategies could include establishing benchmarks in the work program where an assessment is made whether proceeding further is appropriate.[8-5] While at that stage in my career, I hadn't thought about the value of creating formal benchmarks on the Cementon water and sewer projects, I had recognized the futility of continuing on each at a relatively early stage and advised the client accordingly.

My responsibilities at Cementon also included managing the environmental assessment and design of access improvements to accommodate heavy traffic to the plant site during construction. Building the proposed power plant would require over 3000 construction workers in the peak year, and the local road network couldn't handle that volume of worker traffic arriving to and leaving the site.

Our efforts included identifying if anything of archeological significance would be impacted by the access improvements. One day I went to the field with our archeological subconsultant, Karen Hartgen, to understand her efforts, and in a matter of minutes, she found a Native American arrowhead on the site. It was moving to think that arrowhead was made and used hundreds of years ago. Later in the walk, Hartgen found a football-sized rock with a fossil of a sea animal embedded in it; she said the fossil was common and of no archeological significance. Hartgen gave me the rock, which I passed on to the high school Earth Science teacher in the town where I lived.

We knew field crews would walk through areas where there were poisonous snakes and advised them to wear the proper boots to protect themselves. While we warned everyone about poisonous snakes, we failed to mention poison ivy, and two of Hartgen's crew came down with such serious cases they had to go to the hospital — the most unusual worker's comp claims on any of my projects.

When reading the project's environmental reports, I noticed all positive environmental impacts were stated in terms indicating they definitely would happen, while negative impacts were modified with adverbs and adjectives such as possible, maybe, and could. It was apparent our environmental experts let partiality slip into in their findings, probably trying to help the client. After that point, I checked all environmental report results to make sure they were free of such bias.[8-6]

An incident at the Three Mile Island ("TMI") nuclear power plant in Pennsylvania released some radioactive coolant in March 1979, and the federal government issued new rules requiring detailed evacuation plans at all nuclear power plants. PASNY liked the work Parsons Brinckerhoff did for them at Cementon and retained the firm to prepare evacuation plans at their nuclear plants at Indian Point in Westchester County and at Nine Mile Point in Oswego County. I was Project Manager on these assignments that involved developing plans to facilitate evacuation of everyone living or working within ten miles of the nuclear power plant after an incident at the plant. Because this was a new federal requirement, there were no established approaches to perform the work, and we had to create the methodologies. We started by identifying the numerous scenarios during which an evacuation may occur such as workday or weekend; day or night; precipitation (rain or snow) or no precipitation; school in session or not; and the happening of a major event such as a local, college football game. To estimate how long it would take to complete an evacuation for each scenario, we approximated the number of people requiring transportation by bus or medical van, the capacity of the road network, population density, etc. We heard some of our methodologies became benchmarks against which evacuation plans at other plants were measured.

Since all nuclear power plants around the US had to prepare similar evacuation plans, I saw a marketing opportunity. I produced a brochure page on evacuation planning and sent it to the firm's operations managers with nuclear power plants in their area suggesting they contact the plant owners. Subsequently, two of those leads resulted in wins for Parsons Brinckerhoff to provide evacuation planning services.

On evacuation studies, I frequently worked with the client's lawyers. I observed many client actions implied they felt their lawyers were more important than their engineers. On that basis, whenever I saw the potential for a scope increase, I tried to prove the value for the additional effort to a client's lawyers first, rather than to their engineers. My attempts weren't always successful, but I believe I increased the approval rate.

58   THE ENGINEERING IS EASY

While working on various projects in this period, I regularly testified as an expert at hearings and discovered testifying can be a cat-and-mouse exercise. Not only must you know your topic well, you must be prepared for questions from opponents to your position who may state questions in ways to embarrass you or throw you off your game. I saw two incidents of otherwise good and talented people who made poor experts on the stand. One, a brilliant consultant for a planning firm, completely lost his composure on the witness stand when the questions ranged far afield from what he thought he'd be asked. He should have prepared better by assuming he would get tough questions on topics that were peripheral to his basic efforts on the project. Another witness was the manager in an environmental firm who didn't know the specifics of what his staff had done, and thus his testimony turned out to be worthless. He shouldn't have testified as an individual, but on a panel with a member of his staff who knew the details.

When a questioner tried to get me to admit some hypothetical, off-the-wall incident was possible, I learned to say that anything is possible, but that doesn't make it likely. I also never offered my own clarification of a poorly stated question, but rather asked the questioner to restate the question. Some questions consisted of ramblings by the questioner, with multiple questions tossed in haphazardly, similar to this invented question: "Please tell us if the design is safe, because I don't think it is because I don't believe your assumptions are realistic, are they, and you didn't consider all options, for example did you fail to consider a bridge option, something I didn't see in your report; your answer, please."

I loved questions like that because I'd simply respond with a compound answer such as, "Yes, yes, and no," which would throw the questioner off stride as they no longer remembered what they had asked and in what order. I would laugh inside watching the questioner try to regain their composure.

I testified for 15 days at hearings held in a meeting room in the Thruway Motel in Albany on the firm's efforts at the Cementon power plant. Those hearings were held jointly by the U.S. Nuclear Regulatory Commission and the New York State Department of Environmental Conservation, and there were five administrative judges on the bench representing the two

MANAGING MULTIPLE MULTI-DISCIPLINARY PROJECTS   59

agencies. During my second week on the stand, a citizen questioned me after lunch in a very warm hearing room. The questioner would go on endlessly before finally getting to a question he'd frame obtusely about some minor topic. It was all I could do to keep from dozing off before he got to the question. At a point during his rambling, I glanced over and noted four of the five hearing judges were asleep; one judge even had his chair laid back horizontally. I looked to see if PASNY's attorney saw what the judges were doing and realized he was catching up on his mail and not paying attention to the hearing. It made me wonder about the real value of what we were doing.

At the Cementon hearings, I was asked about traffic numbers we prepared. I said I didn't have the back-up with me, but would have an answer the next morning. I intended to ask Al Schaufler, a senior Traffic Engineer in NYC who had developed the numbers, for the information that evening. Schaufler would have left work by the time the day's hearing ended, so I called him at home (no cell phones or texting back then). Schaufler had just moved, but I knew the town and street where he now lived and got his phone number from Information. When I called, a woman answered, and I asked for Al Schaufler. He came to the phone, and I greeted him with my name and started to tell him what I needed. "Who are you?," he interrupted.
 "It's Bruce Podwal, Al."
When he replied, "Do I know you?," I realized something was wrong and asked if he worked for Parsons Brinckerhoff. He said he worked for AT&T, but recently had heard another family named Schaufler had moved in a few houses down. He offered to walk over and have the other Al Schaufler call me back. Soon afterwards, I got a call from the correct Schaufler and everything was resolved in time. However, think about the coincidence it took for things to end well. It turned out Parsons Brinckerhoff's Al Schaufler had an unlisted number, and if there hadn't been another person of the same name on that same street, I never could have reached the right Schaufler in time to get the information I needed to maintain credibility at the hearing.

At the Cementon power plant hearings, I testified for five days alone and ten days with others as part of a panel. Mel Stein served with me on one four-member panel and was a major asset at the hearings. Stein was a Sanitary Engineer by training, and his testimony at the hearings focused on water-related environmental issues. He subsequently became proficient in project controls, and we worked together on many major projects. We also shared a number of similar experiences: Both of us went to CCNY, served as Manager of Project Controls on major projects, were President of the firm's bowling league, and were President of ASCE's Metropolitan Section. Given how stressful testifying under oath can be, it was a comfort having Stein on a panel with me.

My next project was a multi-modal assignment for PASNY who was considering building a coal-fired power plant on Hart Island in Long Island Sound off the coast of the Bronx. Back then, Hart Island mainly consisted of a potter's field for the interment of those who had no other place for burial. Parsons Brinckerhoff was retained to study ways for workers to access the island during construction and operation. I was the Project Manager as we evaluated various access options including a bridge, a fleet of ferries, a tunnel and a tramway. Dave Ozerkis, a CCNY classmate, worked for a governmental agency and was finishing a project management assignment for the Roosevelt Island tramway in NYC. He was very helpful in pointing out issues to address, as tram design was not a strong point at our firm. After evaluating the various modes, the use of ferries became our recommended option, although ultimately Hart Island was eliminated as a site for the plant for environmental reasons.

Hart Island wasn't the only project I worked on that went nowhere. From 1970 to 1977, I started managing the planning and design of many projects, but project after project was cancelled. Because one reason I became a Civil Engineer was to see the public benefitting from my efforts, it was very depressing to work on that many canceled projects. Two projects were cancelled because of economic or financial issues: Westway in New York City; and South Mall Expressway Extension connecting Albany and Rensselaer, NY. The following projects were stopped for environmental reasons: Toms River Extension in Monmouth and Middlesex Counties, NJ terminated because of community opposition; Prattsville (NY) Pumped Storage Plant cancelled because trout fishing might be impacted if a one-in-1000 year drought occurred during construction; and the Cementon nuclear power plant cancelled because the cooling tower plume would be visible ten miles away for several days a year at Olana, the historic home of Hudson River School artist Frederick Church.[12]

I wrote letters very formally back then. For example, I might start a letter with something such as, "We respectfully request approval of our design at your earliest convenience."

Bill Cucolo, a PASNY senior manager and a good friend of Parsons Brinckerhoff, said my letters had to be more specific and less flowery; if I needed an answer back from PASNY by a particular date, I should just say so. After his comments, I made my correspondence to PASNY and others more direct.

---

[12] I had not heard of Olana before working on the Cementon project and decided to visit it. Since that first visit, I've been to Olana several times and highly recommend touring it if you're in the Hudson, NY area.

Chapter 9

MANAGING PROJECTS FOR PRIVATE SECTOR CLIENTS

Project management basics are the same whether working for private sector clients or government agency clients. That said, there are some differences to be aware of, typically relating to money and time, as private clients use their own money or money they borrow and want a quick return on their investment. This chapter uses case studies from five projects with private sector clients to both examine pitfalls and describe lessons I learned.

Shopping Center Developer:

In the late 1970s, I was Project Manager on a project to provide Interstate highway ramps that would connect to a proposed regional shopping center in Pennsylvania. The shopping center developer was the client. It was my first project where the client was a private company and not a government or quasi-government agency. It's where I learned "time" can be a major difference between a private sector client and a public one. While there are some public sector employees who act as if they have no incentive to ever finish a project, time is money to the private sector. To expedite decision-making, the shopping center developer provided private planes to fly me from a general aviation airport near my home to their office and back. Because a plane always was waiting for me, I could stay at meetings until every issue was resolved which eliminated the need for a future meeting. By the way, I've never been comfortable in small planes and always required a co-pilot.

Given the importance of time to them, the client expected us to jump whenever they asked, and working overnight and weekends was frequently required. I quickly realized some of our New York City staff were ineffective in the private sector environment because they were so used to working at a more measured pace for public sector clients. I was forced to request that only those willing to work overtime were assigned to the project. It wasn't an easy requirement to fill as, unlike employees of many other NYC firms who counted on overtime pay, relatively few of our employees back then wanted to work overtime on a regular basis.

Everyone was working flat-out to stay on schedule, as we approached the final submittal due date. It was during that time that we were asked to support a charity that was important to the principal owner of the developer. The charity was holding a black-tie dinner at the Plaza Hotel in NYC. At the dinner, I met a junior principal of the development firm (call him Tommie). Tommie's first words to me weren't, "Hello, how are you," or "I'm pleased you're supporting the charity," or "I hear you guys are working hard to help us get the job done."
No, what he growled at me was, "What're you doing to help me?"
I was taken aback by his obvious attempt to bully me and muttered something perfunctory in reply. Tommie was mistaken if he thought what he said was going to motivate me to do more. I knew how hard everyone was working, and that we were doing it because we had pride in what we do. Engineers work harder of their own volition and respond better to "Thank you" than to intimidation.

I provided expert testimony on our efforts during environmental hearings for the project, and afterwards, PennDOT engineers provided testimony relative to their review of our design of the ramps. The shopping center's attorney asked me to cross examine the state's engineers. I found questioning experts testifying under oath in front of a hearing officer was even more difficult than providing my own testimony; that it's not easy to frame a question to elicit the desired response. At best, my attempt at cross examining an expert rated below average in my opinion. Fortunately, my imitating a lawyer didn't kill the project, as the client ultimately received a permit to build the shopping center.

After we finished the preliminary design for the shopping center ramps, our client asked us for a scope and price to do final design. When we gave it to them, they shopped our scope and price around until they got the cheapest price they could find, and then dumped us. It turned out the firm they selected cut costs by using part-time engineers in the evening who worked full-time for other firms or government agencies during the day, an approach that rarely results in quality products. Oh well, I hope the developer got what they paid for. As the saying goes, "If you think good engineering is expensive, wait until you try cheap engineering."

Real Estate Developer (I):

Parsons Brinckerhoff was asked to perform an engineering study by an investor with interests in Jackson Hole, Wyoming. Most people coming to that area flew into Jackson Hole Airport located inside Grand Teton National Park. However, the National Park Service was considering eliminating any public airport in a national park, including Jackson Hole. Should that airport close, the best alternative to access Jackson Hole by plane would be to fly into an existing airport in Driggs, Idaho and then drive over the Grand Tetons on Route 22. Unfortunately, this road was closed frequently in the winter because of snow, ice and avalanches, and winter naturally was the prime time when skiers came to Jackson Hole. The investor knew closing the Jackson Hole airport would cause local land values to fall unless something was done. Parsons Brinckerhoff's role was to determine how to improve Route 22 such that it would handle the increased traffic and be closed less frequently in winter.

I met with Wyoming's Governor Ed Herschler and received approval to get design criteria from the Wyoming DOT for use in developing concepts for improving Route 22. Next, we estimated the cost and schedule to design and construct the improved road, including building snow sheds to protect drivers from avalanches. We also estimated how many days a year the road would be closed because of weather even with snow sheds. The findings were summarized in a letter report which was used successfully by our client to convince the Park Service to leave the airport as-is. Some 40 years later, I flew out of that airport, reputed to be the only one remaining in a federal national park.

Our fee for the Route 22 study was but $6000 in 1980 dollars. Because this was the only highway project in mountainous regions Parsons Brinckerhoff had worked on in anyone's memory, I prepared a brochure page for the project to highlight the firm's mountainous region experience. For the next decade or so, Parsons Brinckerhoff probably spent twice what it earned on that project reprinting that brochure page as it pursued other highway projects in mountainous areas.

Real Estate Developer (II)

We were completing a heating, ventilating & air conditioning assignment for a Virginia developer who had not paid us for work we had finished. I said we should stop work until the developer both paid us for what we had done and paid us in advance for the balance of our efforts. A partner of our firm said we should finish our assignment as the developer said he'd pay all he owed us when we gave him our final report. I opposed that approach as I felt withholding the final report gave us the most leverage with the developer to be paid. The partner overrode my objection and had the report completed and delivered to the developer — and of course, the developer never paid us another cent.

You may not know why someone didn't pay you for what you did — perhaps they're short of funds for the moment or maybe they're trying to get something for nothing. Regardless, a poor payer in the past, is very likely to be a poor payer going forward.[9-1]

General Public Utilities ("GPU"):

The interview to win an evacuation planning assignment for GPU, the owner of the Three Mile Island nuclear power plant in Middletown, Pennsylvania, turned out to be unusual. GPU asked to meet the proposed Project Manager (me) at the TMI facility at 1:00 pm. I wore a suit and tie, showed up on time and was directed, not to an office building, but to a field trailer. There, the interview team was sitting in flannel shirts, jeans and boots, eating a late lunch of pizza and soda. I immediately realized I was grossly over-dressed and quickly removed my jacket and tie, sat down at the table with them, and grabbed a slice of pizza. From lunch talk, we gradually drifted into how Parsons Brinckerhoff could help them, and a few weeks later, they called to say the firm won the assignment. That interview showed me how important it is to fit into a selection panel's comfort zone.[9-2]

After completing our planning efforts at TMI, I testified in Harrisburg, Pennsylvania in 1981 at the nuclear power plant restart hearings. As the evacuation planning Project Manager, my testimony related only to how long an evacuation would take and not to the basic safety of nuclear power plants. Nevertheless, an opponent of nuclear power approached me after I testified and said he had put a bomb under my car. Even as my pulse rate jumped, I guessed (hoped?) he was bluffing to intimidate me and ignored his threat. After all, if someone really wanted to bomb me, I knew I couldn't stop them. I'm neither brave nor foolish, just a realist. I never was bombed so, thankfully, my guess worked out in the end.

In addition to making threats, some nuclear power opponents were just nauseating. A Parsons Brinckerhoff speaker was obviously pregnant when she presented at a public meeting for a different evacuation planning project. She described the process for determining evacuation times for different scenarios, without taking a position on whether she felt nuclear power was safe or not. I was disgusted by the nuclear power opponents in the audience who yelled out they hoped she'd lose her baby. Of course, most nuclear power opponents aren't shameful. One such opponent worked for Parsons Brinckerhoff and told me he wouldn't work on any of our evacuation planning projects. I said I understood and was comfortable with his decision. Nevertheless, one day he proved he was a team player. When he saw how hard his office-mates were working to complete a major evacuation planning deliverable, he pitched in and helped without being asked.

Casino Owner:

The firm pursued an assignment in 1997 for the program management of the Atlantic City/Brigantine Connector ("AC/BC") project in New Jersey where I would be the Program Manager. Our efforts would include preliminary design, environmental permitting, preparation of design-build bid documents, and administration of final design and construction for a four-mile, $300 million highway including a half-mile tunnel. In the design-build approach, the owner hires a single entity to both design and build the project. In the more traditional American design-bid-build approach, the owner hires one entity to design the project and then hires a second entity to build it. The AC/BC client was a developer who needed the connector to handle traffic to a new hotel/casino complex it was planning; the road's owner would be a quasi-state agency, the South Jersey Transportation Authority. The idea of working for a major casino developer sounded exciting, and I hoped we could win the assignment.

George Friedel was our pursuit manager for the AC/BC project and suggested stressing in our proposal and interview that we understood that casino financials could be in jeopardy if the road cost too much or its opening was delayed. We exhibited sensitivity to those key client concerns by posting two one-word signs behind us at the interview: **COSTS** and **SCHEDULE**. We also showed our willingness to get in the boat with the developer and share their pain and glory by proposing an incentive-disincentive package.[9-3] To wit, our Principal-in-Charge ("PIC") offered to waive 25% of our reimbursement if the project was either over budget or late, provided we receive a bonus of 25% if both costs and schedule were met. The developer accepted the offer and selected us to provide the services.

The developer wanted us to begin work immediately, yet Parsons Brinckerhoff had a myriad of project start-up procedures that had to be completed quickly. These procedures included preparing a project management plan, quality plan, work breakdown structure, and budget. Given the project's size and complexity and the need to start so many tasks simultaneously, I added temporary staff to the team during the project's first month. I asked the local Area Manager, Greg Soriano, to assign someone from the firm's internal Professional Growth Network ("PGN") to the ad-hoc team who he felt had the potential to manage a large project one day. PGN members are employees with limited experience. I felt a major project comes along rarely in most offices, and we shouldn't fail to use its start-up as a training opportunity.[9-4]

The experience I gained working for a shopping center mall developer in Pennsylvania 20 years earlier came in handy as I pushed everyone to stay on budget and schedule. It quickly became evident that most designers were reluctant to think of cost saving approaches, much less evaluating them for incorporation into the design. They constantly had to be prodded to seek ways to develop designs that could be constructed economically. I had to be persistent reminding designers that meeting minimal design standards did not automatically mean a design was unsafe. For example, reducing lane widths by six inches would lower costs by four percent, yet when asked if such lane width reduction met standards, they would say, "You don't really want me to check that, do you?"

When I'd reply, "Yes, I do," they would respond something such as, "No, really?"

This typical dialogue happened repeatedly, as I strove to make sure we delivered a project that came in at or under budget, all of course without compromising our professional judgment. After all, delivering a safe and acceptable project within budget is what the owner, the client, and we all wanted.

The developer's financial model indicated that the road's construction cost could not exceed $200 million[13] or the new casino may not be viable. When we finished the design, our estimate indicated the low bid should be about $200 million which would satisfy the targeted amount. Now, we had to wait for the bid opening to see if the costs were on budget, and we earned the incentive. Three design-build teams were approved to submit a bid. The client asked me to open the three bids in a public forum and announce them to the audience. The first bid opened was for $230 million, which I hoped the developer would say was okay if it eventually was the low bid, even though it was $30 million more than the target. The second bid was for $310 million, which was too much. I assumed the bidder threw in a high bid hoping no other bid was acceptable, and they could walk away with the chance to make a huge profit. There was one bid left to open, and it was for [*drum-roll, please*] $190 million! The price was unquestionably agreeable to the developer, and it meant we earned the 25% incentive. A perfect win-win.

Nevertheless, that year was bitter-sweet for me. Mary was diagnosed with cancer and passed away three months later. At the time of her death, she

---

[13] Dollar amounts in this case study are indicative of the process used, rather than the actual amounts, some of which were confidential.

was an antiques dealer with merchandise in two New Jersey antique centers. I always tried to compartmentalize my business life and my family life and strived never to bring a problem from work home and vice versa. However, creating such a wall was near impossible this time. Fortunately, I had a good team at work, especially Deputy Project Manager, Rich Fischer, and the team covered for me whenever I had to take time off to help Mary during her illness. After she died, I had to close her business while still doing my day job. The closure took two years, working on nights and weekends. Everyone deals with personal tragedies differently. Certainly, both my project work and the effort closing the business were supportive distractions that helped me deal with private, emotional issues. Since that experience, I tried to be more compassionate with staff I knew had a problem at home. Should you be faced with a work-home problem, I suggest you advise your supervisor so he or she can provide support as necessary.[9-5]

While the first phase of the AC/BC project was winding down, I competed as Project Manager for two other projects in New Jersey, hoping Parsons Brinckerhoff would win at least one. The firm won both! Fortunately, the start-up for one of them was delayed a year, and I was able to manage each in turn. When the second phase of the AC/BC project commenced, I turned the Program Manager position over to Fischer and started on the first of the new wins. Winning those projects and staying in New Jersey proved convenient for me as it gave me the opportunity to close Mary's antique business in a measured manner.

CHAPTER 10

MANAGING A MEGA-PROJECT IN NEW YORK CITY

After 13 years managing a broad range of projects, I was appointed to replace Jenny and manage Westway in 1980. Westway, the highway mega-project in New York City, was the second largest active project at Parsons Brinckerhoff, exceeded only by the Metropolitan Atlanta Rapid Transit Authority ("MARTA") transit project the firm was performing in joint venture. It would be the largest project I'd managed by a factor of ten. Little did I comprehend how unprepared I was.

Jenny had been our Project Manager on Westway since it began in 1971. He announced his retirement, as the project's first phase was ending and before the firm's efforts would expand as the project moved to the next phase. We performed Westway in a project office located in a separate building from other firm activities and projects. Several NYSDOT staff and numerous other consultants working on Westway also worked in that project office.

A mega-project is one both much larger in fee and with more interdependencies and interfaces than most other projects. Westway qualified as a mega-project because it would cost over $2 billion to build, Parsons Brinckerhoff's portion of the fee exceeded $40 million in 1980 dollars, and the firm had over 100 full-time staff and twelve subconsultants on the effort at peak. The Manhattan Interstate highway generally would be in a tunnel located slightly offshore from the existing Hudson River shoreline, and fill would be placed over the tunnel for new parkland and development.

When I was designated Westway's Project Manager, there were a few engineers working on the Westway project more senior and more highly compensated than I was. (It reminded me of the modestly-paid university president who is the titular boss over the highly-compensated football coach at many major colleges — and you know who most think is more important.) I feared some of those senior engineers might be upset they hadn't been chosen as Project Manager instead of me, and that I would have to be tough as nails and come down hard on them to make sure they understood I was in charge. It was Mary who counseled me to take a soft approach and start by telling each how much I valued their contribution and looked forward to working with them. Mary's tactic worked, and I never had a problem with anyone.

Regardless, I still had to show everyone I was the right choice. I'd done well managing projects before, so began by employing the micro-management style I'd used successfully for years. It wasn't long before I realized I was drowning. I couldn't keep up with all the meetings I felt I had to attend, quality was suffering, delivery dates were missed, and staff seemed disconnected. I tried everything I could think of, but nothing worked. It made me wonder why I was struggling and heading towards failure, if I was so skilled in project management. I appreciated I needed help, but didn't know whom to ask now that Jenny had retired. I'd been appointed Westway Project Manager by the local geography manager and the project's PIC, neither of whom had ever managed a project anywhere as large and complex as this one. They'd given me general advice so far, but nothing really meaningful as they didn't seem to understand how to manage Westway any more than I did. It was evident to me that if I didn't turn things around quickly, either the client or the PIC would have me replaced.

As far as I knew, the MARTA Program Manager, Jim Lammie, was the only Parsons Brinckerhoff Project Manager who had ever managed anything comparable in scale to what I now was struggling with. I'd never spoken with Lammie, but heard wonderful things about him from my former bridge partner, Howard Chaliff, who had moved to Atlanta to work on MARTA. Chaliff told me how hard a worker Lammie was. He said if Lammie invited you to a seven o'clock meeting, you had to ask if he meant the 7:00 am meeting or the 7:00 pm meeting. Long hours didn't concern me; after all, I had worked with Jenny.

I cold-called Lammie and invited him to come to New York City to do a peer review of Westway. Note a peer review is when an independent individual or team with appropriate skills reviews and critiques what is happening. Lammie knew of Westway and was very willing to spend three days performing a peer review. He sat in on meetings, interviewed project team members one-on-one, spoke with the client and then gave me his findings and advice. Lammie suggested project policies I should institute and modifications to my organization structure. He also recommended I visit MARTA to see how his project was organized and learn the roles and responsibilities assigned to his staff. I made the trip, and it was eye-opening how smoothly everything was running. It was easy to appreciate that many approaches in running a successful transit mega-project were directly applicable to running a successful highway mega-project or a mega-project of any discipline, for that matter.[10-1]

Because of Lammie's advice, I realized I couldn't micro-manage Westway for it to be successful, as a mega-project is too large for one person to manage. I now understood how necessary it was to trust my deputies and associates to share that management burden, and what controls I required in this new paradigm.[10-2] I was still responsible for those I trusted and needed controls to make sure they were doing the right thing. Consequently, I modified the Westway organization structure and instituted new project policies.

As an aside, some years later there was a Project Manager in California (call him Gary) who micro-managed to the extreme. I was PIC on Gary's mega-project, and his staff frequently complained to me about Gary's management style and lack of faith in them. I spoke with Gary several times about this issue, but he was unable to change without reverting to micro-managing at the first sign of a problem. Gary's staff ignored him whenever he said he wasn't going to micro-manage them because they knew he'd revert to his old habits whenever something bothered him. Several of his team said they would leave the firm, rather than work with him. To settle things down, I suggested to Gary he tell his staff that anytime someone catches him micro-managing, they should call him out on it and he'd donate $25 to charity with no questions asked. My suggestion worked for a few months and helped the team get over a rough patch. After that, Gary slipped back into his former style because he was incapable of permanently changing his way of doing things. His failure to change resulted in a few employees leaving the firm, but once Gary quit the firm, those remaining breathed a sigh of relief.

Back on Westway, Lammie also reinforced what Jenny had taught me, namely that schedule and cost control doesn't happen because you want it to happen; you must take positive actions to institute control systems. Using MARTA's approach as a model, I assigned someone on Westway to read every piece of correspondence looking for anything with schedule or cost implications. Then, when something was identified that negatively impacted schedule or cost, I immediately brought it to the client to get them to acknowledge the change's impact and either agree with it or rescind the directive that caused the impact. Obviously, sensing something will happen is more useful than reporting something has happened.

Before long, the Westway project was operating smoothly (or at least as smoothly as a project of that size normally operates), and I finally could de-stress a bit. I had learned that size matters — that complexity increases with a project's size, and what works for small or even large projects may not work for a mega-project.(10-3) I now had become a Program Manager, where program management can be defined as:

> The leadership and day-to-day guidance and control of a complex array of interrelated projects, phases, and activities which must individually and collectively fulfill specified qualitative and quantitative requirements and adhere to schedule and budgetary constraints in order to achieve the overall objectives of the undertaking.[14]

To understand the difference between management and leadership, I suggest reading *The DNA of Leadership* by Dr. Bill Badger. Overly simplifying a bit, management is knowing what has to be done and then making sure everything is done properly, while leadership involves influencing others and visioning what may happen.

The chief Structural Engineer on Westway was Bob Warshaw, and the chief Highway Engineer was Al Manfred, each very experienced in what they did. I'd worked with both many times before, including on I-787. Warshaw had a construction background and the exceptional ability to

---

[14] I received this definition from Lammie.

foresee how a design could actually be built. After every business trip with Warshaw, he always said how much he enjoyed our trip together and how much he felt we accomplished. It made me feel good inside to hear that from someone whose opinion I trusted, even once I realized he probably was saying it by rote. It taught me the value of saying something similar when I traveled with anyone reporting to me. Manfredi, a CCNY alum, had the best talent for visualizing a highway design in three dimensions of anyone I ever met. On Westway, Warshaw and Manfredi would have loud, heated discussions on how to solve a design problem, such that you'd think they hated each other. Then one would notice what time it was, and say to the other, "It's time for lunch, where should we eat today?"

The two good friends then would go off to lunch, before returning afterwards to resume their animated discussions.

The Westway client was NYSDOT, and their Project Manager was Mike Cuddy, who treated Parsons Brinckerhoff firmly, but fairly. After his Westway assignment, he rose to Chief Engineer at NYSDOT. I recruited him to join Parsons Brinckerhoff when he retired from NYSDOT. I typically was wary of hiring long-time, public sector employees, as the shift to the private sector can be difficult, probably as much as for long-time private sector employees who try the public sector. However, having worked with Cuddy on Westway, I felt he would be successful in the private sector because of his technical skills and political astuteness. I was right as he became one of the firm's best operational managers.

Westway's estimated construction cost in 1980 was $2.2 billion, although project opponents claimed it would cost twice as much. Note the cost today would be well over $10 billion in current dollars. NYSDOT had an independent group value engineer the $2.2 billion estimate, and the group found some major savings, including $250 million in foundation cost reductions. Because the group's estimate put our estimating skills in doubt, I checked with my design team for their assessment of the savings. Disappointingly, one engineer said he knew of the $250 million in foundation cost reductions for some time. He was keeping the savings in his back pocket thinking it would make us (mostly him) heroes to identify a cost reduction "at the right time." A talented designer, his judgement in this case was poor as keeping this secret to himself only made us look incompetent when others identified the saving to the client before we did.

I laid-off my first two employees while managing Westway and did it very poorly both times. In the first instance, I never gave the employee any warning a layoff for lack of work was impending, and the whole process was too cold and impersonal. The employee was a senior engineer who deserved much better. The second person I laid-off was an office boy who frequently was late to work. I felt terrible about letting him go because I knew he was poor and needed the money, but believed I shouldn't condone his repeated tardiness. A few months later, it occurred to me that perhaps he was so poor he couldn't afford an alarm clock, and I should at least have asked him if he had one at home. That way, I could have bought him a clock, if he needed one. Once I realized that possibility, I felt even worse about what happened.

After those two incidents, I tried to be more mindful of how my actions affect an employee's life. The more warning I give that a change in employment status is possible, the easier it is for someone to plan their future.[10-4] As soon as I knew I might have to lay off someone because of lack of work, I'd tell that employee that while I'm trying to find work for him or her, I'm not optimistic. I'd then suggest he or she may want to start looking elsewhere for another position. The employee then has the most time to find another job, which is the same courtesy I would want if the situation were reversed.

A technician we hired seemed perfect; he was a go-getter type and worked very hard. Boy, did he have us fooled. One day, someone said the technician came to work waving a pistol around before leaving the office never to return. We then called the police to describe what had occurred. Two days later, a police detective came and said the technician had been on parole when we hired him and would be headed back to prison if the allegations were true. The law typically allows limited background checks before hiring someone, and I recommend availing yourself of such checking to the extent permitted by law.[10-5]

Another lesson I learned the hard way had to do with managing subconsultants. We had contracted with a specialty sub to deliver a technical report on the design criteria for tunnel pumps on Westway. The sub spent one-third the budgeted costs and billed for one-third their fee; we sent the invoice to NYSDOT, and the sub was paid. Next month, the sub billed for and was paid for the second third. Finally, they turned in the report and billed for the remaining amount in their contract. We sent the report to NYSDOT, who rejected it as meaningless fluff and demanded all their money back. At that point, I carefully read the report myself, and while I didn't appreciate every mechanical engineering aspect, I understood why NYSDOT felt the report had no substance and was of little value. I had made a serious error by not assigning one of our own Mechanical Engineers to review and critique the report before sending it to the client. Since then, I allocated money to oversee a subconsultant's work to make sure we give the client what they expect and deserve.[10-6]

The Westway project was the first where I used a sophisticated, computer scheduling system, a newly developed tool back then, to establish the critical path[15] and evaluate progress against the schedule. It took a long while to create the initial project schedule as the process was new to everyone, but once we did, the system poured out an impressive amount of data each month. Fairly quickly, it became obvious we were spending a lot of effort each month developing and inputting data and printing out detailed schedules for various entities, but had little time available to analyze the output before we had to start developing and inputting the next month's data. Knowing we were behind schedule was fine up to a point, but we had no ability to do much about it. We required time to strategize about options to recover schedule slippage in the limited time we had between monthly cycles. With that realization, I made it obligatory for schedulers to be skillful enough to analyze the output quickly and not just technicians who only are capable to input data. Qualified schedulers

---

[15] A critical path is the series of activities which takes the longest duration to perform and sets the shortest time possible to complete the project.

are not easy to find, and over the years, I probably rejected half the schedulers offered to me.

Six months after he peer reviewed Westway, Lammie became my supervisor when he was assigned to New York City to manage the North Atlantic Region consisting of southwestern Connecticut, New York, New Jersey and eastern Pennsylvania. Because I was managing Westway, the largest project by far in his region, Lammie included me on his Regional Executive Committee. On reflection, I think Lammie planned to assess my actions on the committee while developing me as one of his potential successors. Lammie was a former military officer and had been trained to have a successor in mind from Day 1, something everyone should do.

During his Executive Committee meetings, Lammie made sure everyone had a chance to comment on what was under discussion while he developed lists of the pros and cons of alternative actions. At the end of each meeting, he individually asked each participant if they had anything to add. This approach of generating participation showed he truly valued everyone's opinions and suggestions. His style bred loyalty to him and support for the final decision. All of us who worked with Lammie tried our hardest to please him; we never wanted to disappoint him.

I learned from Lammie never to say I'm sorry about a mistake I made, but rather to show what I would do differently the next time I faced a similar situation. Lammie found out what was happening by asking several people the same question; he knew there was a problem when the answers were different. It showed me you learn a lot if you ask the right question and are open to answers that may disappoint you.[10-7] Also, that those secure in their own mind are willing to hear negative things about their own performance, for they know they benefit by having those around them tell how their actions affect what is happening.

CHAPTER 11

MORE PROJECT MANAGEMENT ASSIGNMENTS

Detroit:

In 1993, Parsons Brinckerhoff won the assignment to design a depressed, airport access road at Detroit's Metropolitan Wayne County Airport. The firm's designated Project Manager (call him Phil) waited months for the client, Wayne County, to give the firm notice-to-proceed ("NTP") so the project could start. Believing NTP wasn't forthcoming soon, he retired out of boredom as there was no other work in the office. A month later, the County got their act together, and we finally received NTP. At that point, Phil said he was happy in retirement and didn't want to return to work. No one in Parsons Brinckerhoff's Detroit Office was qualified to be Project Manager, and I agreed to go to Detroit to fill that void until we found someone. Regrettably, Phil, who seemed in excellent health, died only three months after he resigned and never got to enjoy retirement.

The Detroit area office was treated poorly by its regional management headquartered in Chicago. For example, when the Chicago office bought new computers and furniture because its computers were outdated and furniture too rickety, it would send the cast-offs to Detroit and expect Detroit's staff to be grateful. Also, to keep rent costs low, regional management sited the Detroit office in a run-down neighborhood in a 30-story building that was over half empty and surrounded by other mostly empty, tall office buildings. That part of town was so deserted that I joked I could cross the street blindfolded in mid-day and not worry about being hit by a car. Given the office's furnishings and location, it was embarrassing when clients visited. The only decent place I could find to eat lunch was a Black Muslim vegan restaurant two blocks from the office. I still recall their great soups, salads, and sweet potato pie.

While I was disappointed in the office and settings, I realized a Project Manager can't always be choosy as there are only so many projects out there. Sometimes you get a good posting and other times a not so good one. That said, I enjoyed working with the Detroit staff, found the project

MORE PROJECT MANAGEMENT ASSIGNMENTS

interesting, and liked the assignment.(11-1) I especially remember my staring contests with the hawk who often sat on the fire escape outside my office window. It was easy to become licensed to practice engineering in Michigan as it had a reciprocal agreement with New York, where I had passed my original test to be licensed; I merely had to prepare an application and pay a fee.

The most complicated design aspect of the project related to elevated ramps connecting the airport access road to a proposed new terminal. The ramp structures required long spans of relatively shallow structural depth on very sharp horizontal curves. None of the Detroit Office engineers had ever designed a similar bridge before. Not wanting to risk a design mistake, I described my concern to the Regional Manager and asked his help finding someone qualified to design thin, sharply curved structures. A week later, the most senior Bridge Engineer in Chicago arrived in Detroit. I asked if he had ever designed a similar structure before, and he responded, "No, but I'm sure excited about the opportunity to do so."
I thanked him and immediately sent him back to Chicago. Then I went out of region to find someone at Parsons Brinckerhoff with the required experience and was pleased with the engineer who was available. The Regional Manager was understandably upset his region wouldn't be getting labor credit for the work as that was a metric he was evaluated on.

While I realize I was senior enough to reject someone I felt unqualified without fear of retribution, I encourage even new Project Managers to verify the capabilities of every key person assigned to their project.(11-2) The message here is don't struggle with unqualified staff on your project. Get the level of expertise required to do it right the first time. It's one thing for a junior person to learn on-the-job while working under a qualified senior person in their discipline; it's another when the senior-level person isn't qualified and intends to learn on-the-job. Training to overcome deficiencies of senior personnel is their supervisor's responsibility and should be covered by their overhead budget, not by a project's budget.

Whenever anyone senior I didn't know was proposed to work on my project, I reviewed their résumé and asked to see their last two performance evaluations. If their supervisor told me that I wasn't allowed to see the performance evaluations, I'd say, "In that case, I reject the candidate."

And before you knew it, they sent me those evaluations, and I was able to make an informed decision as to whether that person was suitable or not. Further, whenever the best internal person available still wasn't the right person, I went outside the firm to find that right person. In most firms, the policy is Project Managers have total responsibility for the success or failure of projects. If that's truly the case, then Project Managers must have the right to reject unqualified project personnel. I also used that right to move anyone off the project who may have had the qualifications, but was not a team player. Note I didn't mind someone disagreeing with me on various issues. It's when someone was unhappy or unwilling to work on the team or with me, that I saw little reason to make him or her do so.

I had been working on the airport access road for several months when the Michigan Department of Transportation advertised for a firm to do the design and environmental impact assessment for widening nine miles of I-94, an Interstate highway in Detroit. The project included a major investment study ("MIS")[16] and extensive public involvement efforts. Alternatives included carpool lanes (also referred to as high occupancy vehicle, or HOV, lanes), light or heavy rail in the median, and exclusive busways. The construction cost was estimated at $1 billion. As the first phase of the airport project had ended, I felt I could turn that project over to my deputy, if we won the I-94 project with me as the Project Manager.

The Regional Manager, based in Chicago, designated himself as PIC on the I-94 pursuit as he wanted a major role in the interview. Sensibly, he didn't resist when I said the client knew he wasn't based in Detroit and was more concerned who would be the Project Manager and what the Project Manager's skills were. As a result, we increased my face time with the selection panel at the interview. During the interview rehearsal, we made sure everyone knew how many minutes they were allocated to speak, so my time to sell myself as Project Manager to the client would be sufficient. However, at the interview, our Traffic Engineer got on a roll

---

[16] A Major Investment Study is a tool to improve metropolitan area transportation planning at an earlier time than previous methods.

and didn't stop talking. Fortunately, I was close enough to be able to kick him lightly under the table, and he quickly wrapped up his presentation.

One of the project tasks was public involvement, and a subconsultant suggested a women-owned business (call it PS Inc.) that they had worked with previously to perform that task and that would help us meet our affirmative action goal. We added PS Inc. to the team, but I was very disappointed with the proposal text they prepared and what they wanted to say at the interview. It became evident they were public relations specialists, not public involvement, and had no knowledge of the public involvement process required in the environmental regulations. I had to rewrite much of their proposal text and their interview presentation. In any case, we won the assignment, almost in spite of them.

Selection for the I-94 project was my fifth consecutive winning interview as Project Manager. Not just after a loss, but also after a win, I always requested a client debriefing. At the I-94 project debriefing, I asked what the client liked about our proposal and presentation and what they liked about our competitors'. The client said one of the other teams had a much better public involvement firm than we did. Because I wasn't impressed with PS Inc., I asked if he would be supportive if I added the firm he liked to my team, and he said he would. We still would satisfy our affirmative action goals as the new firm also was a women-owned business. Thus, as soon as PS Inc. finished their basic efforts, I ended their role on the project and gave the new firm the additional tasks that evolved as the project progressed. I saw no reason to struggle with a poor subconsultant, similar to my rejecting an unqualified engineer on the airport access road.

We were required to include an estimate of hours necessary to perform the scope. At the debriefing, I asked if our estimate was reasonable when compared with those of the other firms that submitted a proposal. When told our estimate of hours fell in the middle, it gave me leverage when we were later pressured to reduce our costs to perform the work.

The Regional Manager had contacted a major competitor of Parsons Brinckerhoff and offered them a subconsultant role and the Deputy Project Manager slot to induce them to join the team pursuing the I-94 project. When work started, I realized it was awkward for the deputy to be from a subconsultant when Parsons Brinckerhoff held the prime contract with the client. For example, when I was away from the office on vacation, I didn't want the deputy to agree to a client request that had contractual implications of which the deputy had no appreciation. After that project, I never gave the Deputy Project Manager slot to a subconsultant.[11-3]

I hadn't thought I was capable of racial insensitivity, but an incident in Detroit showed I was naïve in that regard. On June 17, 1994, the office conference room TV was on, and several people were watching O.J. Simpson's white Bronco being pursued by the police in slow motion on the California highways, after his ex-wife and a man had been murdered. Seeing there was no way for OJ to escape, I said, "What's he hoping to do? Doesn't he know they'll catch him eventually?"
Immediately, the three African-Americans in the room turned around and glared at me. It quickly became obvious to me that they interpreted my comment to mean I, a white man, had already assumed the black man was guilty of killing the white woman and man. It made me aware we all have to be careful or racial insensitivity can slip into our everyday thoughts and actions.[11-4]

While working in Detroit, I generally stayed weekdays in a furnished apartment in Southfield, Michigan and returned home to New Jersey on weekends. Whenever I was assigned outside the NY/NJ area, my wife had to hire contractors to perform routine household tasks I normally would have done, including snow shoveling. The New Jersey snowstorms were especially harsh the winters I was in Detroit, and the cost for snow removal became the tipping point in the decision to sell our house and move to a rental apartment. An example of how the benefits from a career of working worldwide come with a price to one's family.

Buffalo:

After a year, the I-94 project no longer was a high priority to the client, and the environmental process stalled as the client wasn't reviewing submittals in a timely fashion. As the work slowed, and I was no longer full-time billable on the I-94 project, it was propitious for me to find

MORE PROJECT MANAGEMENT ASSIGNMENTS

another project. An opportunity arose in Buffalo, and we began developing Parsons Brinckerhoff's number two person on the I-94 project to replace me if we won the Buffalo project. The potential new project was the Southtowns Connector/Buffalo Outer Harbor project consisting of twelve miles of highway, light rail or busway alternatives from Buffalo southward. The project scope included a design study, an environmental impact assessment, an MIS, and extensive community involvement. Cost of the contemplated improvements was estimated at over $500 million.

Our interview was scheduled for the Buffalo Hyatt ballroom at 11:00 am, and we would be the second team presenting that day. We rented a meeting room in the Hyatt so we could assemble the morning of the interview and go over last minute instructions. A subconsultant asked if they could bring an outside railroad consultant, call him Nick, to the interview to answer any questions on railroad coordination As I had a rule that no one attends an interview who hasn't been briefed about their role, I replied Nick can attend, provided I meet him prior to the interview and advise him on what to do at the interview. I said Nick should come to our meeting room at 9:00 am on the day of the interview. Nick showed up almost an hour late and said he had trouble finding us as he had gone to the wrong room. He appeared somewhat befuddled, but I told him not to worry as there was time for me to brief him, which I quickly did.

I hadn't worked in Buffalo since college, and the prospective client, the local NYSDOT office, didn't know me. While I understood the environmental and MIS processes from my work on the I-94 project, I felt the NYSDOT interview panel would doubt I knew the corridor's local issues. To counter that concern, we started the interview presentation with a video of a flyover of the corridor. While the video ran, I explained what we were looking at and described the key issues in each area. We also had retained and videotaped a local TV newscaster, Sandy White, doing one-on-one interviews with four elected officials from Erie County and the City of Buffalo and with a community group leader opposed to the project. Presenting that second video at the interview reinforced the fact we understood local issues and concerns.[11-5] By the time the two videos ended, I was confident NYSDOT felt my team and I knew both the corridor and its issues.

The interview was one of the smoothest I ever participated in, and we were notified a few weeks later that we won the job. However, a major problem surfaced before we started contract negotiations. It seemed the wrong room Nick had gone to on the interview day was the Hyatt ballroom. The room was dark because the interview of the first team was underway, and that team was giving their slide presentation. Nick assumed it was our team practicing, but when the lights came on and Q&A started, he became confused when he couldn't spot anyone he knew. After a while, Nick had realized this was not the right team, and he left the ballroom and found us in our meeting room.

The team, whose presentation Nick blundered into, challenged our selection, claiming we gained an unfair edge because Nick heard some of NYSDOT's questions. NYSDOT was obliged to conduct an investigation including interviewing Nick and me. After the investigation, NYSDOT was satisfied Parsons Brinckerhoff neither gained an advantage nor had done anything wrong and affirmed our firm's selection. Need I mention I never used Nick on the project? Obviously, had Nick explained to me what had happened at the beginning, I might have been able to resolve the situation before it became serious.

With the win, Parsons Brinckerhoff opened an area office in Buffalo. I was asked to be the Area Manager, in addition to being Project Manager, but said I preferred to focus on managing the project. Jeanine Viscount ultimately was chosen as Area Manager, which meant I was reporting to someone far less senior than I was. I was fully comfortable with that arrangement because I could focus on what was important to me: the project. By the way, Viscount did an excellent job in Buffalo, her first Area Manager role.

This was the second project I worked on that involved an MIS, still a relatively new requirement for federally funded highway projects. Greg Benz was Parsons Brinckerhoff's MIS expert and had received a contract from the federal government to hold a series of seminars on the principles of MIS for government employees around the country. I taught the design portion of the MIS process at the seminar held in Newark, NJ.

A project purpose was to encourage development along the Southtowns corridor. I spoke with a Parsons Brinckerhoff land-use planner to get his sense of how a highway project could facilitate development. He said development is often a zero-sum game in a region that isn't growing. If a developer builds something in such a corridor, it's likely an existing development somewhere else in the region will suffer. For example, a new mall will likely take business away from the existing malls in the area. The discussion left me feeling that it's likely some statements about how a project encourages development are based more on faith than facts.

The NYSDOT understood they should fund our efforts at $15 million for the first three years, but they only had $5 million to allocate for that time frame. Thus, the project moved very slowly from its beginning as it never was funded to the required extent. We had fourteen subconsultants scheduled to work on the project, and I had to tell seven they would have no work for at least three years because of the lack of funds.

Soon after starting, it became apparent the state, county and city all had widely different expectations of what they wanted the project to accomplish. We tried everything we could think of to build some sort of consensus out of the numerous stakeholder workshops, but nothing worked. It even proved impossible to reach agreement about eliminating obviously inadequate alternatives from further consideration. The project underscored that when various governmental entities have different agendas, the simplest of decisions may take months or even years.[11-6]

With the project slowing down, it didn't make sense for me to waste project money treading water, and I agreed to pursue another project management assignment.[11-7] When Parsons Brinckerhoff won that project, it was easy to convince NYSDOT to accept my Deputy Project Manager on Southtowns, Dale Moeller, as a qualified replacement with the added benefit of being a cost-saving measure. Note, it's over 20 years since I left the Southtowns project, and only a small portion of the originally contemplated improvements has ever been constructed.

CHAPTER 12

MANAGING AN ASSET MANAGEMENT PROJECT IN NEW YORK CITY

In 1989, Parsons Brinckerhoff pursued a primarily vertical (buildings) project in New York City, where I would be the firm's Project Manager if we were successful. The project scope involved performing condition assessments and developing maintenance programs for every capital asset worth more than $10 million (in 1989 dollars) that New York City owned. The purpose for a condition assessment was to determine if the asset was in good working order, and if not, to establish what had to be done to get the asset back to good working order. There were over 2,000 such assets, and they included such diverse facilities as Yankee Stadium, the Brooklyn Bridge, the Metropolitan Museum of Art, prison ships, piers, incinerators, hospitals, and schools. The contracting agency was the New York City Office of Management and Budget ("OMB").

A buildings project would be new to me, but not my largest hurdle. I would also have to learn how to incorporate rapidly evolving, hardware and software technologies into our efforts, and we would have to add many architects to our office, which had only one. Further, the work had to be completed in only one year. This combination of factors (new technologies, new client, new employees and tight schedule) involved the potential for large risks especially related to quality control.

We heard another firm pursuing the assignment was the front-runner, and we tried to think of ways to impress the client. One of our IT staff mentioned there now was a 9"x12" computer that you could hold in the crook of your arm [*keep in mind this was 1989*], and he wondered if that may be something to use on the project. While a computer is just a tool, I immediately visualized how we would do things quicker with fewer errors by using portable computers and how such a computer could be a "wow-factor" selling point to highlight in our presentation. I had a computer programmed as we would use it doing assessments, and at the interview, I walked up to the selection panel to show them how we intended to perform the work. I also offered to train city personnel on the computers so they could do future annual updates themselves. The strategy worked, and we won the assignment with me as Project Manager.

Even before receiving notice that we won the assignment, the manufacturer had stopped making the computer I used at the interview because competitors already were making much smaller (4"x10"), lighter and more powerful models that could be held in the hand. We contacted three firms manufacturing hand-held computers and asked for a price for the 24 computers we would require to do the work. A concern I had related to the durability of the hand-held computers because it was likely our staff occasionally would drop a computer while in the field. I stated bidders had to include a loaner computer with their bid and that we would test durability by dropping the computer onto a concrete surface from a four-foot height. Serendipitously, the lowest priced computer also passed the durability test, weighed the least, had the longest battery life, and had the most memory; it was our clear-cut computer of choice.

We created a master program whose output provided a ten-year estimate of how much funding for capital and operations & maintenance was required to keep each asset in working order. Because we had to finish in one year, I came up with a concept whereby different teams would perform their tasks simultaneously rather than in series. Every office and field team then could input findings independently into the master program, and no team had to wait for another team to finish before inputting data.

Most field teams were a group of three: A Mechanical Engineer, an Electrical Engineer and a Registered Architect. Both to save time and avoid transposition errors, field staff had to enter data directly into the hand-held computer, rather than first writing it down and later entering it into the computer. All input was menu-driven, and staff couldn't finish a site assessment until they answered every question on the menu. Most staff didn't realize there was a time-clock in the computer, and we could tell when they input the data. The time-clock reassured us that staff weren't sitting in a restaurant all day and then entering invented data at close of business.

We noted one Electrical Engineer (call him Terry) entered the data each evening at around 7:00 pm. From his team leader, we found out Terry was with the rest of the team at the site every day. When we asked Terry about the data entry issue, he admitted he was uncomfortable with computers and pretended to input data into the computer in the field, while surreptitiously making notes on paper he carried. When he got home, he gave his teenage son the notes to enter into the computer after dinner. We told him that approach was unacceptable, and suggested he transfer to another project where he'd be more comfortable. Terry agreed, and we replaced him with a different Electrical Engineer.

Given the condition assessments were primarily subjective, I knew some individuals would evaluate condition significantly different than others would. For example, an asset I rate as in good condition, you may assess is in fair condition. As a quality check, I scheduled for randomly selected assets to be checked a second time by a different team so we could compare assessments. By that approach, we identified individuals who were either too conservative or not conservative enough and coached them how to evaluate future assessments consistent with the evaluations by others.[12-1]

The quality check approach proved its worth when a serious personnel issue arose. All the architects on the project were new hires, and one of the requirements was that they were Registered Architects. States were not yet posting lists of RAs online, and it took our Human Resources Department a few weeks to verify if all new hires were registered. We discovered one new hire (call him Don) was neither registered nor even a college graduate as he claimed, and we had to let him go. Don had completed several asset inspections to that point, and I sent a Registered Architect to reassess the assets at our own expense. The reassessment confirmed the initial findings by Don which showed our quality control process was so effective, we were able to train a relatively unqualified individual to produce satisfactory results.

I accompanied some teams on a few asset visits to determine if the approach was going as intended. Carnegie Hall was one asset I still remember visiting as we inspected it. For a few moments, I broke away from our inspection team and sat in the balcony of a nearly deserted hall listening to an orchestra rehearsing. It was heavenly.

Half-way through the OMB project, I was asked to help chase and win another project that would require me to relocate cross-country. When we won the assignment, I turned the OMB project over to Mike Rugless, the Deputy Project Manager on that project. The project was going smoothly, and the client was pleased with Rugless's efforts and agreed to the change. Ultimately, our entire approach on the OMB project proved successful, and we completed the effort on time. The client was pleased, and we continued providing updates for the next three years.

Some Project Managers start a project and feel obligated to stay to the end. However, while I would have liked to have finished the OMB assignment, I accepted that a Project Manager's responsibility also is to help win that next project. If I could turn a project over to a new Project Manager at the mid-point of effort, I could try to win the next project before finishing my current one and then go on to repeat the cycle over and over.

What I've typically found is the thorniest phase of a project (and the most interesting to me) is the first third. That's where:

- We learn how different the project we've won is from the project we believed we were chasing,
- We find out if the client treats us the way we thought they would,
- We create and implement the project policies and systems, and
- Staff sort out their roles on the project.

The second third of projects is often routine maintenance of the procedures and systems put in place during the first third, while the last third typically involves a gradual winding down of effort and staff.

After I've set everything in motion in the first third of the project, I want to give my successor, usually my Deputy Project Manager, the opportunity to become Project Manager and enhance his/her résumé should I win and start my next project. If I properly trained and developed the deputy, the client is willing to let that person become Project Manager, knowing the deputy is capable of finishing the project (and probably costs less than I do). When a client preferred I stay involved after turning the project over to a replacement, I would continue on the project as a technical advisor or PIC. In that situation, I would juggle my workload so I could both provide oversight of the previous project and start the new one.

Note we were able to use our experience from the OMB project to win future assignments from other clients in this market area. My former bowling teammate, Yalcin Tarhan, was designated the firm's lead in this area, which became an important market for us. My last involvement with asset management came 25 years later (four years after I'd retired), when I moderated a session on transportation asset management at the 2015 ASCE annual convention in New York City.

CHAPTER 13

MANAGING OPERATIONS

Lammie appointed me to replace him as Regional Manager when he became Parsons Brinckerhoff CEO in 1983. I always felt my asking for help on Westway, before ordered to do so, was one reason why Lammie chose me to replace him. A year later, my region was expanded (to be called the Atlantic Region) to include Delaware, Maryland, DC, Virginia and western Pennsylvania, and it now consisted of twelve area offices.

Regional Manager was my first non-technical, non-project related assignment, and I had to learn on-the-job how to apply project management and leadership skills to an operational assignment. My only guide to being Regional Manager was what I'd seen Lammie do, so I tried to replicate him. Complicating matters, I hadn't read one of the management books that discuss how difficult it is to replace a beloved leader and offer suggestions on how to succeed in that role.

Lammie's selecting me for the role dispelled personal doubts that I was qualified. Still, I knew I had to prove I deserved to be in charge and could make the decisions required to be successful. I worried if I could fulfill the objective metrics for an operations manager: (1) keep overhead costs on budget, (2) make the margin (profit) goals for projects, (3) collect invoices in a timely manner, and (4) win new assignments. A major task to meet those objectives involved managing the region's 80 project managers. I was confident I could provide supervisorial oversight of project managers as I would have wanted if the situation was reversed.

I'd seen the benefits of participatory management working with Lammie, so I often asked my two regional deputies, Paul Byrne and Julie Hoover, for their opinions before I made decisions. Byrne was excellent at damage control; when a client was upset at us, he was the one I sent to calm the client down and recommend an approach going forward. Hoover was the quintessential example of a manager who cared for the welfare of her people and was as an effective advocate for their needs.

92 THE ENGINEERING IS EASY

Early in my tenure as Regional Manager, I asked Byrne and Hoover for their opinion about a decision I planned to make on a minor matter. We discussed the pros and cons, and both recommended strongly against my position. Nevertheless, I said I was going ahead with it anyway. However, after going home and thinking about what I planned to do, I realized I only had asked Byrne and Hoover for their opinion because I thought they would support me. I now understood their points against the decision were good ones, and also that ignoring their input meant I'd lose their future support on issues more important than today's minor one. Next day, I called them into my office and said since they both felt so strongly, I'd rethought my decision and decided to defer it for now.

To me, walking the walk on participatory management meant encouraging those with whom I worked to suggest solutions and ideas and then treating their recommendations with dignity. My goal was to be a confident leader, clearly in control, yet flexible enough to accept opposing positions presented for the common good. Only then, could I benefit from participatory management.

There were a number of joint venture ("JV") projects in my region, where the JV partner was a frequent Parsons Brinckerhoff competitor. Too often, staff from both JV firms carried that competitiveness over to the project, and there was little teamwork happening. As a board member on several JV boards, I felt it incumbent to make sure project staff realized I opposed in-fighting between the firms. While those at Parsons Brinckerhoff battled hard when competing, we expected them to team civilly in a JV. Parsons Brinckerhoff's policy was to treat other firms the way it would want to be treated, even when other firms didn't reciprocate.

During this period, three officers of Parsons Brinckerhoff left to form their own firm in Baltimore. We didn't want to lose them, but wished them well. However, soon after starting the firm, they began poaching our clients and staff. It was also clear they were trading on proprietary information they gained while they had worked with us. While some of my co-workers wanted me to go to the clients they took from us and offer to do the work for less money until we drove their firm under, I knew that wasn't the Parsons Brinckerhoff style. I'm proud I took the ethical approach.

One morning, I was having coffee in a Pennsylvania diner with a major transit client's Project Manager (call him Wayne), who was overseeing our efforts on the client's largest design project. Wayne mentioned he had managed another project where his agency self-performed the design. A design error was discovered during construction on that project, and a portion of newly constructed rail tracks had to be removed and rebuilt. Naturally, the contractor submitted a claim for additional compensation to remove and rebuild the tracks. Wayne said he didn't want to ask for more money from his supervisor because of a design error on his watch, so he told the contractor to do the extra work for free, or he'd make the contractor's life miserable. Wayne proudly said the contractor acquiesced. My heart sank as he told this story. All I could think was that one day he would treat us as poorly as he treated the contractor. I warned our Project Manager to be prepared to push back against Wayne's unprincipled ways, but we failed to do enough. We ultimately lost quite a bit on the project, when we couldn't document and prove we deserved additional reimbursement for our claims.

Because interest rates were high in the mid-1980s, clients slowed payments owed consultants so they could keep funds in their bank accounts as long as possible. At the same time, Parsons Brinckerhoff was in a tight cash flow position, and billings and collections were a major concern. In fact, several times the firm worried if it could make payroll. Fortunately, our CFO, Allan Stevenson, was able to convince the banks to help keep Parsons Brinckerhoff afloat until it collected monies clients owed the firm. There were hundreds of active projects in my region, but the largest dozen resulted in over half the region's billings, somewhat analogous to the premise that 20% of anything causes 80% of the problems. I focused efforts on collecting those dozen invoices, and before long, the region's cash flow improved significantly. It's about determining one's priorities and then dealing with the highest priority first — Dyckman's Theory of Limited Objectives.

If someone I worked with disappointed me, I tried to deal rationally with the situation. I avoided being angry at them (especially in public), as that only might make them too nervous to perform well, or in the extreme, they might even consider how to sabotage me. Whenever someone made an error in judgement, I used that as a teaching opportunity so they wouldn't make that error again. If, even after additional coaching, I felt they were incapable of doing better, I limited the complexity of assignments I gave them going forward. When someone did something really egregious, such as lying or being unethical or abusive, I'd resolve not to work with that person anymore. The next two examples show how I dealt with unacceptable actions by senior-level employees.

We were a subconsultant on a New York hydroelectric project and our client, the prime consultant, was based in Boston. I was going to Boston to attend a conference of a professional organization and thought it worthwhile to make a courtesy call on the client while in town. Before leaving for Boston, I asked our New York-based senior Project Manager how the project was doing, and he replied everything was going well. When I walked in to meet the client's Project Manager in Boston, he immediately said how glad he was to see me. It seemed we were having major problems on the project, and he assumed I had come to resolve those issues. Because I had no idea we had any problems, I, of course, came with no solutions to offer. I did what I could on the fly and got out of there as soon as I could do so gracefully. When I got back to New York, I asked our Project Manager why he never warned me what I would face and let me be blindsided. His mumbled excuse was insufficient, and I gathered he had been afraid to admit that there was a major issue on his project. To me, it was unacceptable for a senior manager to cover up a problem such that I was put on the defensive with a client. I was inwardly furious and looked for the next opportunity to replace him. I always tell people that if I hear about a problem from you first, I'll help you solve the problem with no recriminations.[13-1] However, there's no get-out-of-jail-free card when I first hear about the problem from someone else — certainly not when someone else is the client, and the offender is a senior manager who should know better.

MANAGING OPERATIONS

One Area Manager had a habit of giving people nicknames that he used in private and not to their face. Now, funny and maybe even mildly insulting, nicknames between friends and family are one thing, but he gave nicknames to clients that I felt were childish, offensive and very inappropriate. I told him to stop, if only because a mocking name eventually would get back to a client who may not see any humor in it. He never stopped using nicknames, and it counted against his judgment in my opinion; I felt I had no choice and began the process to replace him.

Obtaining environmental approval for Westway, still the largest regional project, was dragging on, and in 1985, the project came to a premature end. New York was concerned it might lose the project's federal funding if Congress didn't extend the nation's Interstate Highway program's end date, and the state decided to trade the project's federal highway funds for transit funds. Without Westway, two other mega-projects (MARTA and Pittsburgh Light Rail Transit) now sustained the firm, while all the rest of the firm's projects combined generated a loss. What that implied to me was we should use small projects to train and develop staff and as entries to key clients, but that without large projects, it's unlikely the firm will prosper. Meanwhile, I was aware that the unexpected, sudden end of any large project has a disproportionate impact on the firm's health.

Any premise of avoiding small projects was tested when a young new hire in the Traffic Engineering Department (call him Hank) said his buddy, who worked for Hartz Mountain a New Jersey developer, offered to retain us for $500 to do a few days of traffic engineering work. I rejected the idea saying it costs more than $500 to put a project in our system — in other words, avoid this small project. However, Hank's Department Head, Chuck Kole, said Hank was so enthusiastic and motivated that couldn't I please agree in this case. I reluctantly said okay, but asked Kole to keep his eye on everything. Several years later we finished our last assignment for Hartz Mountain. By then, they had paid us $6 million in fees (in 1980 dollars), which shows how fortunate it was that Kole convinced me not to turn down that small project.

The firm was providing services to Paterson, New Jersey for a proposed, hydroelectric facility at the Great Falls[17]. Our task was to develop the bid documents so Paterson could advertise for a developer to finance, design, build, operate and maintain the facility. We were nearly complete, when Paterson said they had no money to pay us to complete the work. However, they said if we finished, they would reimburse us when a developer came on board and paid them a development fee. Alternatively, if no developer came forward, we never would be paid. We thought the project had such good potential that we agreed to finish the work at risk, providing we got paid with back interest of 6% should a developer come forward. Paterson agreed to our offer in writing. Four years after we finished our work, a developer came on-board. At that point, we requested payment of our principal plus the interest, which by now had grown to 25% of the principal. Paterson, which had forgotten about our principal much less any interest, refused to pay us anything and, in fact, threatened us with a counter claim. After some contentious negotiations, we settled for less than we were owed, once we determined we wouldn't do better given the cost of defending the counter claim.

One lesson from Paterson is your value decreases with time when clients somewhat conveniently forget all the hard work you did and any commitment they made. Think carefully about providing work products for a client if you know payment will be significantly delayed. A second lesson is, if you decide to proceed anyway, send the client a monthly statement (every month for four years in Paterson's case) so they have no excuse to claim they forgot they owe you money.

George Alexandridis, with whom I worked at Brill Engineering, contacted me. He was leaving an assignment as CEO of Lockwood, Kessler & Bartlett, a mid-size engineering firm on Long Island, NY and was checking to see if I had a spot for him at Parsons Brinckerhoff. Alexandridis had been slightly senior to me at BEC, and back then, I recognized he had great potential. Because Paul Byrne was returning to project work, I was able to appoint Alexandridis as a Deputy Regional

---

[17] Alexander Hamilton had used the falls to power a manufacturing plant in the early 1800s.

Manager. It never bothered me that, given his talents and experience, Alexandridis might one day become my supervisor (which he did four years later), as his joining the firm could only make us better. As the old saying goes, "A-level players surround themselves with A-level players, and B-level players surround themselves with B-level players."[13-2]

My secretary booked me into a four-star hotel for a business trip, because she had heard good things about it. The hotel was very nice, but on checking in I was surprised to see the cost per night was over $200 at a time when we usually spent half that amount. I felt guilty spending so much, and to offset the hotel cost, I decided the right thing was to charge my labor for the trip to vacation, rather than an administrative overhead account. Two weeks later, Lammie called me in to say he reviewed my expense report and wondered why I stayed in such an expensive hotel. He was satisfied when he heard what I had done to offset the cost and approved my expenditure. I saved myself a major embarrassment by doing the right thing of my own volition.

Business Development:

Winning projects means more work for everyone, so I learned all I could to be good at business development. Business development in a consultant firm involves such components as identifying clients and markets to pursue, writing proposals, attending interviews, and negotiating agreements after selection. I became fairly good at most components, other than visiting prospective clients I'd never met to see if they had a possible assignment for us in disciplines that weren't my strength. Rather than completely avoiding such clients, I usually brought someone knowledgeable in that discipline to keep the discussion moving.

Most of what I learned about business development came by observing the Parsons Brinckerhoff marketing managers in the '80s. Collectively, they were the envy of our competitors, as they always seemed to know with whom to team and how to structure winning proposals and presentations. I also read numerous books and articles on business development, where the best book I read on the subject was *Marketing Architectural and Engineering Services* by *Weld Coxe*. Each book and article on the topic had all sorts of tips, such as always do this or never do that, which were somewhat similar author to author. However, I now was learning about those grey areas when a tip shouldn't be followed blindly. For example, say we sense we're in third place going after a project. It's logical to assume the two firms ahead of us know the universal marketing tips, which means following the tips is unlikely to vault us into first place. To paraphrase Ralph Waldo Emerson, "Consistency is the hobgoblin of little minds, philosophers and kings," which says to me that we can't do the same thing as everyone else and assume we'll be successful and those ahead of us won't. We must find a major game changer to increase the chance of winning or else save our money and look elsewhere for work. Successful game changers I utilized over the years included:

1. Hiring a superstar to propose as the Project Manager when the available Project Managers on staff aren't that outstanding.
2. Adding a world-class consultant firm to the team that can address selection panel concerns that may be raised at the interview.
3. Altering the client's perception of what it needs, sometimes referred to as changing the playing field. E.g., convincing the client that mining skills are more necessary than tunneling skills when we have better mining capability than our competition.
4. Doing something dramatic at the presentation, such as illustrating a concept by preparing a design sketch in front of the interviewers.

The above game changers are more than just smoke-and-mirrors to dazzle a selection panel. They're examples of ways to prove you are offering something of real value.

In the early 1980s, Parsons Brinckerhoff invariably joint ventured with Gibbs & Hill ("G&H") when chasing transit projects in the New York City area. G&H had more local capabilities in the required technical areas than we did and usually provided the JV Project Manager, while we were assigned the routine efforts. The JV team frequently won when we pursued projects with G&H in the lead, but I could see Parsons Brinckerhoff would always be in the back seat with no opportunity to grow. My options were to do nothing and continue to get half the work or take a chance and separate from G&H to see if we could grow our own market. In cliché terms, should Parsons Brinckerhoff settle for the bird in hand or go for the two or more in the bush?

The next time a transit client requested a proposal, I took the risk and brought on board a superstar Mechanical Engineer, with the requisite skills to be proposed Project Manager. I then told G&H we would go it alone. I'm not one who normally takes risks (risk avoidance is my default setting), so I didn't sleep the night before the interview, as I questioned if I did the right thing. Fortunately, it turned out well as Parsons Brinckerhoff won the assignment, and that project became the first step for the firm to become a transit powerhouse in the New York City area. In the next two years, the firm won two more major transit projects with the same superstar as our proposed Project Manager. The game changer worked wonderfully.

For years, I avoided pursuing work from the various New York City public agencies because they sent the message that cost was more important than value. The agencies bid engineering services seeking the lowest cost rather than the most qualified firm and negotiated with consultants confrontationally rather than as equals. A former Parsons Brinckerhoff employee (call him Ike) now worked for the City and encouraged me to pursue a small project he was managing. The project was to upgrade the Staten Island ferry terminal with reimbursement by lump sum. I told him why we avoided chasing City work, but Ike said he would treat us fairly as he knew we would do a good job. I agreed to pursue the project on that basis, but never mentioned I would use that project as a test case to see if the City had changed. I told my staff to estimate our costs and that I would cut our fee to make sure we submitted the lowest price and won the assignment, even if it meant we would lose money. The staff estimated our costs at $60,000 and we bid it at $45,000, which turned out to be low enough to win. Thus, we began processing the contract knowing we'd lose $15,000, which I felt was a reasonable investment to find out if the City had improved.

However, before signing the contract, worrisome things started happening. First, the City changed the project from lump sum to "cost reimbursable," meaning we would spend extra money preparing detailed invoices, further increasing our negative margin. After a few more weeks, the City said they would limit the overhead rate, which meant we couldn't invoice for our full overhead, again increasing negative margin. Next, Ike said he was promoted and wouldn't be the client's Project Manager when the project started, which I felt added to the likelihood we wouldn't be treated fairly. The final straw was when the City's attorney revised the liability clause to the proposed agreement, increasing our liability and the potential of being sued. The costs we already expended preparing the proposal and negotiating the agreement were sufficient for me to decide there was no reason to lose another $15,000, and probably much more, on this project. We'd confirmed the fact the City hadn't changed and still wasn't a good client for us. I called Ike and said we're withdrawing from the selection because of all the changes since we were selected.

Parsons Brinckerhoff's proposed Project Manager was very nervous during interview rehearsals in the pursuit of a Grand Central Terminal transit project in New York City. He wasn't any calmer when we arrived at the interview room. Hoping to settle him down, I suggested he remove his suit jacket and do the presentation in shirtsleeves. It seemed to liberate him immediately, and he was completely relaxed during the formal presentation and following Q&A. Parsons Brinckerhoff won the assignment, which I don't think would have happened if he hadn't been so well composed at the interview.

Many firms always ask for a debriefing from the potential client when the firm lost a project it pursued. That way, the firm may learn what it has to do next time to be successful. Not surprising, some clients are reluctant to be honest about a firm's weaknesses and use platitudes rather than meaningful comments. It's the task of those sent to the debriefing to get the client to open up and share their real feelings, so the firm fully benefits from the debriefing. Parsons Brinckerhoff had lost a Virginia bridge design project it was pursuing, where our proposed Project Manager (call him Hal) previously had managed a bridge project for the same client. I sent George Friedel (my marketing manager) and Hal to the debriefing.

MANAGING OPERATIONS

After introductions, the client began discussing the reasons they selected another firm for the assignment. They started by mentioning that while Hal was a very talented Bridge Engineer, they had a problem with him as Project Manager because he was too argumentative. Hal leapt to his feet shouting, "I am not!," ending any chance the client would provide another meaningful comment at the debriefing.

Parsons Brinckerhoff won an assignment for New Jersey Transit, which required our Project Manager, Sal Matina, to spend significant time in Newark, in Essex County, New Jersey. I took that opportunity to open an area office in Newark, rather than just a project office, and appointed Matina as Area Manager[18]. Matina always asked my approval before making a major decision. However, if I didn't answer him promptly, he took action. I liked that he did something and didn't stop and wait on me, as inaction usually is the worst action.

The day we signed the lease for the Newark office space, I called the Essex County Engineer and told him about our new office. He said the County Executive would be pleased to know we're opening an office in the county and that he would send us a Request for Proposal for a bridge improvement project they recently advertised. We wound up winning the project, which shows one way a firm can leverage new offices.

Human Relations, Training & Development:

Lammie's years in the military involved significant periods of formal training. Because he regarded continuing education as extremely important, he commenced the practice of the firm providing funding to train senior managers. I was able to take an outstanding seminar on how to manage a professional services firm, such as an engineering, architectural or accounting firm. Taught by a combined Harvard-MIT faculty team, it involved seven days of intensive training primarily using case studies. Many case studies dealt with topics covered in a typical MBA program. While I certainly didn't qualify for an MBA after this short

---

[18] Staff in a project office work only on the one project in the office, while staff in an area office can pursue and perform multiple projects.

course, I now had a better handle on how firms operated. The total experience of subject matter, instructors, and classmates was very uplifting, and I stayed on a high for months afterwards.

Four of the seminar case studies were so pertinent to what we were doing that I used versions of them to train key staff in my region. Lessons from the seminar that stayed with me include:

1. Accept the blame when you appoint someone unqualified to fill a slot and they fail; it's not their fault you appointed someone unqualified. Rather, it's your responsibility to spend the time to check your staff's performance, and to help as necessary, until satisfied they are qualified for the assignment.[13-3]
2. When person after person fails in the same position, consider changing the organizational structure as it may be inhibiting success.[13-4]
3. People in large, well-equipped home offices rarely appreciate the hardships faced by those in smaller field or branch offices, which often are stretched thin in staff and equipment. Anyone overseeing smaller offices from afar should spend sufficient time observing those offices and helping as necessary.[13-5]
4. Don't assume everyone acts the same in similar situations; people act differently when they're motivated and evaluated differently.[13-6] Generalizing somewhat for an engineering firm, marketers are optimistic, designers are cautious, and construction engineers are rigid:
    a. Marketers, who are evaluated on sales, tell potential clients, "We can design anything you want at any schedule for any cost," as they strive to win more work,
    b. Designers, who are evaluated on the quality of their designs, say, "Give us more time and unlimited budget, and the product will be even better" as they seek to avoid design errors, and
    c. Construction engineers say, "Simplify the design" as they seek to shorten the schedule and reduce costs; two items on which they're typically evaluated.

Lesson 4 suggests there always will be those who act only for themselves, and that you must be prepared to react to their motivations. However, Lesson 4 doesn't mean people shouldn't act for the common good. It sends a powerful message when those who report to you see you value team players and are one yourself. You want your team to realize they'll benefit when the whole organization does well.

The Trenton, NJ Area Manager was dissatisfied in an engineer's performance and planned to lay him off. However, just before the Area Manager took action, the engineer sent me a letter. In the letter, he asserted his designs were being rejected by other designers in the office who were not qualified to evaluate his work, and those designers also provided unsafe designs to clients. While I sensed the engineer only wrote the letter in an attempt to save his job, I treated it seriously because his assertions could have been true. First, I told the Area Manager to take no action on the engineer. Second, I asked Tom Keusel, a well-respected, senior officer, to interview the engineer and the other designers and to report back to me on his findings. After his review, Keusel told me the engineer's assertions were unfounded, and I reported that finding back to the engineer. I also advised the Area Manager he could take whatever action he deemed appropriate, and he laid-off the engineer.

Up the Organization: How to Stop the Corporation from Stifling People and Strangling Profits by Robert C. Townsend, then President of Avis Rent-a-Car, influenced me greatly during this period. One of his basic tips that stayed with me is that you shouldn't create a form you haven't tested on the firm's President. If you think the President would be reluctant to fill in the form, it's a good indication you realize the form is more complicated and onerous than it needs to be. After reading that book, I always completed any form or survey I created before imposing it on others.[13-7]

I also read *To Engineer is Human: The Role of Failure in Successful Design* by Henry Petroski, where he talks about how we learn more from failure than success. For example, say someone develops a new bridge concept and designs a 100-foot span bridge that is built and performs perfectly. The next thing that will happen is someone will use that concept and design a

200-foot span bridge and then a 300-foot span and so on, until one day a bridge collapses because the design concept doesn't work for that length span. Only after a bridge fails do we have something to study and learn as we try to find out whether a design concept is acceptable.

Feminism had become more established in the 1980s, and there were a number of books and articles written to help women in business. I read several and found one book to be excellent not just for women, but anyone in the business world. The book is *Dearest Amanda* by Eliza G.C. Collins, and it provides first-rate tips on how to maintain a successful work-life balance. While perhaps a little simplistic by today's standards, the book was useful to me for many years.

Sexual harassment claims were just starting to be raised at that time. One day, I heard from a senior female engineer that another female engineer in Baltimore (call her Kay) alleged that a senior male engineer, whom she didn't want to name, sexually harassed her on a business trip. In this instance, Kay was more comfortable going to another female, than to Kay's direct supervisor, a male. I was told Kay didn't want to make a big deal about it. I was relieved to hear Kay wasn't pushing us to do anything and decided to respect Kay's request. Nonetheless, Kay left the firm soon afterwards. In hindsight, I should have had Human Resources talk with Kay as soon as I heard about the situation. Perhaps, I could have saved her for Parsons Brinckerhoff if, instead of marginalizing the incident by doing nothing, I showed her the firm always investigated allegations of harassment and dealt decisively when they proved true. While I was more sensitive to gender issues than I had been 20 years earlier with Martha, I still hadn't progressed far enough as confirmed by Kay's situation. I was learning from my gender-related mistakes too slowly, which hurt my ability to perform.

I did better, human relations-wise, with a talented, senior engineer (call him Larry) with a drinking problem. Almost everyone knew Larry drank heavily and was not to be disturbed after lunch. The first time I evaluated Larry's performance, I read his prior performance evaluations and was surprised to see no one had documented having spoken to him about his drinking problem. After I discussed the situation with Human Resources, they talked with Larry and told him he had no choice but to enter a substance abuse program for which we would pay. He resisted briefly, but then entered the program. It was successful, although a year later, Larry left the firm because of other health reasons.

In addition to Larry, I inherited other employees with issues no one wanted to deal with before. Usually, I feel everyone can provide some value, even if they're overpaid for what they do. I believe we can find ways to use them to their strengths while holding back on salary increases and promotions until their value and compensation are aligned. Lammie had showed one way to compare value and cost was to rank everyone according to their value. The most valuable person was the one I would take with me if I was starting a new operation; the second most valuable would be the next one I'd take, and so on. I then plotted everyone on a graph where their value was the x-axis, and salary the y-axis. I next drew a band of a best-fit curve connecting the dots, and it generally became evident who were the over- and underachievers.

Using that process, three underachieving individuals stood out. The projected workload at that time indicated we had too many people on board, and it was imperative that we reduce staff quickly. Given the situation, I did what had to be done and told the three they were laid off because of a combination of lack of work and below average performance. Two additional reasons convinced me to ask each to leave. First, for years almost everyone knew the three were the worst performers in their respective disciplines. Keeping them sent the message I tolerate poor performance. Second, when the turnaround eventually came and the workload increased, it should be easy to find significantly better replacements for each, raising our level of competence.

A year later, a Drainage Engineer in NYC (call him Jay) faced lay-off as we had no foreseeable work for him. Jay who had worked for us for several years, agreed to transfer to Phoenix if we won an Arizona project we were chasing. It took eight months from the start of the pursuit, through ultimately winning the project and signing the agreement before we received notice to start work. We had paid Jay's salary for that entire time, even though we had no work for him in that period. When told work was starting and he should relocate to Phoenix, Jay said he decided he wouldn't move to Arizona. I then told Jay's supervisor to lay him off. His supervisor suggested we give Jay two weeks' notice and two weeks' severance pay, but I refused, saying we've given him eight months, and you should tell him to leave today. I was very disappointed for two reasons: first, because we kept our part of the bargain and Jay hadn't, and second, I had let down the Phoenix Project Manager who had counted on us to provide a Drainage Engineer.

My secretary in 1984 (call her Liz) sat outside my office, where I couldn't see her from my desk. One day someone said Liz was asleep at her desk, and he had to wake her to ask if he could see me. After he left, I called Liz into my office and asked if everything was all right. I said someone saw her sleeping, and I was concerned she might not be feeling well. Liz said she had not been asleep, and I decided not to pursue it further. However, several days later, someone else came in and told me Liz was sleeping at her desk. Afterwards, I asked her again if she had a health issue that I needed to make an accommodation for. Liz was furious at me, saying this was all an excuse to push her out. I tried to say that all I wanted to do was help her, but Liz left the office in a huff and quit. Fortunately, I'd documented everything contemporaneously, and my notes were used to refute her claim for damages for unjust actions that caused her to resign.

Every so often, someone working for me would complain it was unfair that so-and-so was paid more than they were. My set reply was always to ask if they would be happy if I reduced so-and-so's salary. Next, I would say they have to realize that there always will be some people who are paid more than they're worth, but that after a few years of small or no salary increases, I hope to get their pay in line with their value. Whenever someone came to me saying they were leaving the firm for a higher paying job, I would offer a change in current responsibilities if I felt they were someone I wanted to keep, but I wouldn't look to match the pay increase they would get. After all, almost anyone can get some other firm to pay them extra to entice them to leave, and I saw little benefit winning a bidding war for an employee who already has shown a willingness to look elsewhere for monetary satisfaction. By the way, the tactic that worked for me when I felt I was under-paid was to ask my supervisor, "What do I have to do to earn more money?"

There were many times when I counseled someone unhappy with their supervisor or their position in the firm. Typically, I would say they should be patient for another year or so, and it's likely they'll be pleased as Parsons Brinckerhoff is a good organization where cream ultimately rises to the top; in fact, one day they may find their current supervisor is reporting to them. I'd suggest they treat their supervisor to lunch to settle things down, as I found that usually worked for me when I had an issue with a supervisor (assuming your organization allows you to pay for a supervisor's meal). When the concern was with their place in the organization, I'd mention that most firms constantly change over time. If a firm is centralized, the next time upper management changes, the new leaders will say they have a wonderful idea, and decentralize. The leaders who follow them then say they have a wonderful idea, and they centralize. After all, that's all there is — an organization either becomes more centralized or more decentralized. Your goal is to figure out how you best fit into either organizational structure and seek those positions.

Before becoming Regional Manager, I frequently needed something from the firm's Purchasing Manager. I often felt the he only helped when he had a mind to do so. However, once I became Regional Manager and had some semblance of power, almost overnight, he went out of his way to assist me. I never forgot how he treated me before I was important, and I vowed to treat everyone equally regardless of their level.[13-8]

When visiting an office in the region, I usually held a brown bag session for the staff on some topic. I especially focused on training and counseling the dozens of Project Managers in the region. Developing the firm's future leaders was rewarding, so being Regional Manager was satisfying on the one hand. However, it was not as fulfilling as managing my own project.

Another reason I didn't enjoy being Regional Manager was time management had become a major problem. The number of documents arriving in my in-basket each day was overwhelming. (Time management isn't easier today, as too many people send e-mails to too many recipients without assessing the value of mass mailings.) I felt obliged to read every document sent me, because it might contain something important. What I often discovered was documents were sent by people trying to impress me with what they had done, and not because they needed my input. I used every moment I had catching up on my reading; if I knew I'd be standing on line for five minutes, I'd bring something to read so my time wasn't wasted. My objective of visiting the twelve area offices once a month further limited my free time. My stress level stayed high, as I was unable to do all that had to be done at a reasonable pace. Normally, I try to avoid stressful situations; I say, half in jest, I prefer passing stress on to others rather than taking it on myself.

Time for a Change:

My first year as Regional Manager was a good one financially, and my target numbers for the next year were increased for new business won and project profits and reduced for office overhead and accounts receivable. I had a second good year, and my numbers were made more challenging again. I was competing against myself! In my third year as Regional Manager, I was drained spending time tracking numbers and reading the numerous documents sent to me. Although my numbers were good, it was an uneven three years for me, because I wasn't enjoying myself. I missed project work; it wasn't fulfilling to be two or more levels removed from those working on projects. While I had pride in the position I held and its real and perceived power, I was left wondering when things would improve for me personally.

Solving my angst began in 1986, when Lammie, based on recent growth, decentralized and subdivided a few regions, including splitting the Atlantic Region into four smaller regions. (Of course, some years later, the firm centralized once again.) There were twelve Area Managers in the Atlantic Region, and four, who clearly were ready to advance, became Regional Managers. I moved laterally to become Lammie's staff assistant, where I reviewed projects and operations on a firm-wide basis and learned how to apply both project and regional operational skills at the corporate level.

CHAPTER 14

PROJECT REVIEWS

I frequently was asked to review someone else's project — sometimes by the Project Manager and sometimes by their supervisor. In addition to peer and technical reviews, I provided project start-up help on many projects.

This chapter describes typical project management issues and demonstrates why performing project reviews made me a better Project Manager. Every review reminded me of things I should be doing on my projects or, worse, made me realize there were bad habits I was falling into. Although I didn't serve as Project Manager on any project mentioned in this chapter, they all added to my résumé, and some gave me the personal satisfaction of working on a world-class project.

Reviewing Projects with Significant Margin Erosion:

One of our best Bridge Engineers, Lou Silano, was running out of work in 1984 at a time when there were four major bridge proposals coming up, each in a different part of the country. We designated Silano on the four projects as our proposed Project Manager, hoping to win at least one of the four projects (our normal success rate was about 35%). We were ecstatic when Silano and the four pursuit teams did such a good job at the interviews, that we won all four. Because Silano couldn't manage the four projects simultaneously, other, less skillful Bridge Engineers/Project Managers took control over three of the projects. The euphoria of four wins blinded us to the risks we faced. We failed to protect against those risks, and by the time the four projects had been completed in 1987, combined losses on the four ran in the multimillion dollar range.

Silano's situation was an example where outstanding business development skills were outpacing technical resources. The firm had been winning so much work it was hiring and expanding too quickly; our ability to manage growing operations and perform quality work was stretched to the breaking point. Lammie asked me to determine the specific reasons why so many projects were losing money. I looked at eleven projects whose margin had eroded significantly hoping to find the root causes of their margin erosion (i.e., reduction in anticipated profit). To avoid pinpointing simplistic reasons for margin erosion, such as automatically blaming the client or poor project management, I applied the systems analysis approach I learned some years back.

My findings established that the eleven projects often had more than one root cause for margin erosion, and many shared the same causes. Little did I realize the process would result in a set of margin erosion causes that would guide me the balance of my career.

<u>Margin Erosion Root Causes:</u>

1. <u>Inadequate Budget</u>: In eight of the eleven projects, the project team overspent the budget performing the work because either the firm shaved the price too much to win a bidding war or the scope was so vague, the client could claim everything they wanted was included in the fee. Budget overruns were inevitable in both those scenarios. This root cause reminded me of my first money-losing project, a New Jersey toll road project. In that instance, failure to adequately define the project in the contract led to the client compelling us to design a more complex project with no increase in fee.

2. <u>Unfamiliarity with Client</u>: In seven of the eleven projects, the project team overspent the budget learning the procedures and standards of a new client or one it hadn't worked for recently.

3. <u>Inadequate Skill Sets</u>: In seven of the eleven projects, key staff on the project team lacked the skills to perform the work properly. Thus, the team had to redo poorly performed efforts for no additional reimbursement.

4. <u>Unqualified Project Manager</u>: In six of the eleven projects, the firm appointed someone as Project Manager who had never managed a project of this type before, and the Project Manager's inexperience led to problems.

5. <u>Poor Subconsultant Oversight</u>: In five of the eleven projects, the project team didn't provide proper oversight of subcontractor performance and had to spend its own money correcting subcontractor errors.

6. <u>Improper Project Charges</u>: In three of the eleven projects, overhead costs, such as staff training, were improperly assigned to the projects. Applying overhead costs to the projects made it appear that the projects did worse financially than they really had.

As one looks at the four most prevalent causes (Root Causes 1 through 4), it's apparent someone made poor choices or was overly confident even before work started. Senior management, rather than Project Managers, typically made those choices which generated the root causes. For example, a regional or office manager selected the client, negotiated the contract, and chose the Project Manager and key team members. The senior managers may have been talented, but were too optimistic and hopeful things somehow would work out in the end.

When there is more than one such root cause at a project's start, senior and project management can predict a project is a strong candidate for significant margin erosion.[14-1] In fact [*cliché alert*], one can say three such causes create the likelihood for a perfect margin erosion storm. In such cases, it's prudent to budget extra oversight to lessen the risk potential. Yet, for example, can we expect inexperienced Project Managers to realize they are a high risk on their own project and budget accordingly?

Jump ahead 20 years, and I was teaching an internal, three-hour session in Dallas to Project Managers on preserving margin and controlling margin erosion. I planned to use the six root causes for margin erosion to illustrate potential financial risks and describe actions to mitigate those risks. The agenda had me spending five minutes at the start going over definitions for such terms as margin, margin erosion, revenues, costs, and write-offs[19], but I spent 40 minutes on definitions as most of the Project Managers were clueless about the meanings of the terms. It made me

---

[19] A write-off is a project cost that can't be billed to a client.

wonder how we can fault Project Managers for margin erosion when many of them don't understand what the term means, probably because they're more technically skilled than managerial. Problems occur when someone in upper management creates definitions for terms such as margin erosion and then blindly assumes everyone else properly understands those terms.

Many in upper management have never managed more than one or two basic projects, and thus have little appreciation for the day-to-day pressures on a Project Manager. Still, upper management continues to designate unqualified staff as Project Managers and then hypocritically blames the staff they appointed when margin erosion occurs. It's understandable why upper management is concerned about margin erosion. However, the proper approach is to acknowledge blame when unqualified staff fail at a task to which we assigned them. We must be willing to accept responsibility for appointing staff before they are ready and for not providing them with the support they require to be successful.

Also, senior management should accept that there always will be projects experiencing margin erosion, and as overall revenues increase, it is likely gross margin erosion will increase. To me, a key metric is whether margin erosion as a percentage of revenues is increasing or decreasing.

Peer Reviews:

I reviewed a New Jersey project that had several margin erosion issues. The most significant reason for margin erosion was the failure of the inexperienced Project Manager to document in writing scope changes the client's Project Manager orally authorized (Root Cause 4, above). Thus, when the client's Project Manager became ill and soon thereafter passed away, we couldn't prove we were authorized to do extra work for additional compensation. While we claimed the work was out of scope, the client's replacement Project Manager stated the work was in scope, and his interpretation prevailed as we lacked documentation to contest his ruling. I prepared a case study of the project, changing the proper nouns for anonymity. As I visited offices around the firm, I presented the case study to emphasize good practices to follow. I probably shouldn't have been surprised, but four different Project Managers in four different offices asked how I heard enough about their project to use it to create the case study. What that said to me was whatever is going wrong in one office on one project is happening on many other projects worldwide.

Confucius told us to study the past if we want to define the future. We must learn what mistakes others made and then do something positive to avoid making the same mistake ourselves. Because we're capable isn't sufficient to keep us from making an error, or from repeating an error we knew others had made. After all, we must accept that good intentions aren't a system. Whenever Jenny heard of a mistake or error, he added a new control or system to make sure it never happened again. I often said a good engineer or manager doesn't repeat a mistake they previously made, and an outstanding one doesn't repeat a mistake they know someone else made.

In the '70s, Duttenhoeffer was PIC on a power project in Louisiana. Another engineer and I were going to New Orleans to take a training course, and Duttenhoeffer asked us to spend a day reviewing the project being performed in an office 30 minutes from the training venue. A week beforehand, I contacted the Project Manager, call him Brad, to tell him we would visit the office for the review. Brad said someone would pick us up at our hotel after the morning rush hour ended and drive us to his office. When we reached the office, we found out Brad and his key staff were tied up at a meeting in the field and wouldn't return until lunchtime. Brad arrived at noon and took us for a long lunch during which he gave us a brief overview of the project. After lunch, there wasn't much time to do any sort of a decent review, before we had to leave for the airport to catch our plane. We'd been played by Brad who made sure we didn't have the time to find anything to criticize on his project. Brad knew we were leaving mid-afternoon and was able to orchestrate our schedule to keep us from reviewing anything in depth. I told Duttenhoeffer it was probable that Brad (best case) didn't accept that we were there to help or (worst case) had something to hide. After that experience, I avoided telling reviewees when I would depart, so I could control the review agenda.

Parsons Brinckerhoff was in JV with Bechtel on the Central Artery (a $15 billion new highway), program management assignment in Boston, and one of our firm's largest projects. In 1993, the JV brought in a new Project Manager and Deputy Project Manager, and Lammie, who was a JV board member, requested someone from Bechtel and I perform a change-of-command audit of the project.[14-2] Together, we interviewed 80 project staff and the client to get a sense of what was working well and what wasn't and reported our findings to the JV board and the new Project Manager. One observation was that there was a good working relationship between JV staff from the two firms. In fact, we often couldn't identify the employing firm of those we interviewed, as many seemed only concerned about how well the JV was doing rather than their direct employer. The JV board and the project's previous management team had done an excellent job overcoming the normal tension between employees of two firms who sometimes compete against each other for work.

I was appointed to the firm's Quality Management Committee ("QMC"), chaired by Vijay Chandra. QMC members randomly selected large projects to visit, assessed how the projects were doing, and assisted Project Managers in completing the projects successfully. We didn't look at projects that were in trouble, as those generally already had more help than they could use. Instead, we visited projects that seemed to be going well to see if we could spot future issues no one realized they might face. The QMC's approach was not to point blame, but to suggest solutions to the Project Manager on how to address things with a potential to go awry.

I highly recommend Project Managers review other projects every so often. It's logical that whatever is going wrong on one project is likely to be happening on others, and by reviewing projects, you have the opportunity to learn from the mistakes and misfortunes of others.[14-3] While most Project Managers are reporting at the 20% completion point they'll make the entire positive margin anticipated when the project commenced, somewhere along the way, many projects end with margin erosion. For that reason, you should ask for a review of your own project at the 20% point. The time to control margin erosion is before it happens, not after it's obvious there will be erosion.[14-4] The next four paragraphs depict problems and solutions from four different projects I reviewed for the QMC that no doubt are applicable to other projects.

I was peer reviewing a project in Nevada, not knowing the project had stalled over a design concept. The Project Manager, call him Pete, had proposed a foundation design concept the client liked, while Pete's supervisor, call him Ron, wanted a very different concept. Progress was held up for months as Pete was uncomfortable to bring Ron's concept to the client, because he knew the client preferred the original concept, and was afraid to challenge Ron by proving Ron's approach had problems. The project might have stayed stalled indefinitely if it hadn't randomly been selected for a QMC peer review. I wasn't qualified to review the design concept, so I brought in a foundation expert, Rube Samuels, from New York. After interviewing Pete, Ron and the client, Samuels stated he preferred Pete's concept. At that point, Ron acquiesced, and the project finally got moving again. The lesson is to reach out to an independent expert when you need an impartial arbitrator to break a stalemate.[14-5]

Another project I peer reviewed involved preparing several bid packages for a highway in Delaware with major drainage outfalls. Construction had started on the first design package before I began reviewing the other packages. The Project Manager, call him Lee, said he had retained an independent consultant (a contract employee) to design a major drainage outfall in that first package. Because Lee couldn't find anyone local at Parsons Brinckerhoff to review the consultant's design, the bid package went out with the drainage design unchecked. I feared the contractor would start building something and then have to demolish it because of a design error discovered during construction. I mentioned my concern to Lee, who stated he shouldn't be blamed if something went wrong because he tried to find a local reviewer. I replied that, as Project Manager, he was responsible for everything on the project and should not have stopped until he found someone, internally or externally, to check the design.[14-6] I then identified a Parsons Brinckerhoff drainage design expert from a different office to review the design as a late review is better than hoping there's no error. In this instance, everything checked out fine, but it's important for a Project Manager to have that confirmation in hand.

During a peer review of a New Mexico highway project, I asked the Project Manager and the PIC what worried them the most. They said they were concerned our Geotechnical Engineers may have missed something when characterizing the soils, as the client hadn't authorized all the borings we recommended. They knew the client wouldn't accept any responsibility if a side slope failed during construction, and we would be blamed. The situation brought back memories of a project in Turkey when we were not able to take any borings, yet were blamed for mischaracterizing the rock condition when I failed to document our client's directions. (The Turkish incident is discussed in more detail in subsequent Chapter 16.) Having learned from my mistake in Turkey, I suggested they write the client and recommend assigning a full-time Geotechnical Engineer to the site during excavation, so side slope adjustments could be made promptly when necessary. By this approach, we would be on more solid ground [*another intended pun*] to limit liability should the client fail to take our advice and a side slope slip.

On another project, the Project Manager was leading a three-party JV on a mega-project in Michigan of over 100 staff and spent much of his time either working in his closed-door office or meeting with the client. Thus, most project staff rarely saw or interacted with the Project Manager. My peer review noted morale was lagging, in part because the average person on the project felt disconnected to the project's upper management. The project was accomplishing some wonderful things, but few realized that fact. I suggested the Project Manager hold all-hands, monthly meetings to tell project personnel what had happened and what was going to happen, as a means of building pride in everyone working on the project.[14-7] He commenced holding such meetings, and I saw morale had improved when I returned to do follow-up peer reviews. In some of those peer reviews, I led a team that included members from the other two JV firms to help reinforce the objectivity of the findings.

ACEC Firm-Wide Peer Reviews:

I was active in the American Council of Engineering Companies ("ACEC"), where I served on the national Transportation Committee. As a member service, ACEC facilitated evaluations of an engineering firm's general management, project management, business development, quality, financial management, human resources and computer systems by peer reviewers from other firms. I served on twelve peer review teams and was

PROJECT REVIEWS 117

team leader of eight of those teams. I peer reviewed firms ranging in size from 60 to 1500 employees, and one firm was based in Canada. Often, we reviewed not just a firm's main office, but also its field and branch offices.

I observed that problems in one engineering firm often occurred in other peer reviewed firms. It made me realize that every engineering firm likely has similar problems somewhere in their operations. For example, because several reviewed firms had employees who said their supervisors asked them to mischarge on their time sheets, it's probable that's also happening in many non-reviewed firms. (Note a purposeful time sheet mischarge is a way to cook the books to look good to your boss or, worse, may be an indication of fraudulent overbilling of a client.) My peer review experience indicated there are bad apples in all organizations, and every firm must be vigilant about finding and eliminating them.[14-8]

Another thing I noticed from the ACEC peer reviews is that policy compliance often is watered down for each layer one goes lower in an organization's structure. Management creates policies, circulates them throughout the firm and then frequently sits back expecting the policies will be followed. Well, it doesn't always happen as senior executives assume it will. Employees don't always remember or understand every policy, and if they do, they can be very creative at finding ways to circumvent them. It's not uncommon for an employee to say we always follow a specific policy. However, when prodded, that same employee will mention the situations when that particular policy was sidestepped. The lesson is upper management must constantly confirm policies are being followed by going down into the organization to check.[14-9]

Summarizing my peer review experiences, the following issues surfaced in peer-reviewed firm after firm and office after office, and it's reasonable to assume those same issues are happening in most organizations:

1. <u>Over-Focus on Billability</u>: An over-focus on billability (the percentage of total employee hours spent working on projects, rather than on overhead) to the point it's the key metric, may encourage people to do things they shouldn't. For example, staff may fudge time sheets to look good (e.g., claiming to work on a project when they really are

working on an overhead task). Further, even when their projects are profitable, Project Managers should not permit overhead staff to falsely charge time to their projects.

2. <u>Selecting Unqualified Project Managers</u>: Too many Project Managers are selected because of their technical skills, not their project management skills. For example, discipline experts are designated as Project Managers to win a project, because most clients prefer the consultant firm's best (say) Bridge Engineer to be Project Manager on a bridge project. In fact, supervisors may force some employees to become Project Managers who want to stay technical. In my opinion, technical excellence is not the primary skill set we should require of Project Managers. Rather, a consulting firm should want Project Managers who (a) fulfill the contract terms such that satisfied clients pay on time and want to rehire the firm in the future; and (b) do the work within the budget. I certainly understand there will be instances when a firm may have no choice but to select a technical expert to be the Project Manager to win a project. In that instance, a qualified Project Manager should be designated as deputy to assure project management tasks are performed properly.

3. <u>Poor Quality Control</u>: Everyone says they believe in quality, but there's almost universal admittance they sometimes send out unchecked work because there's no time to do otherwise. Simply stated, they fail to either budget enough hours to provide quality or schedule enough time to perform it. The risk of poor quality is obvious, and positive actions are required to reinforce a commitment to good quality.

4. <u>Poor Risk Management</u>: If one knows about a potential risk at the start, there's no reason to ignore that risk during project execution. However, from top to bottom, confidence causes people to discount risks once a project commences.[14-10] As a minimum, diligent follow-up during project execution is required, and senior managers must work with Project Managers to assess if risk management is adequate.

5. <u>Unethical Practices</u>: People frequently encounter situations that tempt them to act unethically or illegally. Such situations are ubiquitous such

as staff working on a money-losing project being asked to assign their labor costs to a different, more profitable, project. Another example is a construction inspector offered gifts or favors from the contractor to overlook construction defects. Firms and Project Managers must continually assess performance to confirm every one of their staff is doing the right thing and acting properly.[14-11] Further, firms must stress they expect employees to report infractions they sense others may be doing, say by use of a confidential hotline, with no retribution to a whistleblower.

As an example of someone who always acts honestly, I refer to my best friend, Harry Rosenthal, an engineer I met in grad school at Poly. He is the most ethical and trustworthy person I know. Rosenthal worked in an organization where he reviewed contracts for a public-sector owner; he noticed some contracts had an appearance of potential corruption (say, giving work to a firm because of political connections, rather than professional qualifications and/or best price). Rosenthal frequently had to resist pressure from above to approve contracts, until he completed his due diligence and was satisfied everything was acceptable.

Another example is Professor Marc Edwards of Virginia Tech who helped bring drinking water crises in Washington, D.C. and Flint, Michigan to the public's attention. He had to overcome both public officials, community activists and engineering scientists who questioned his professionalism when he said drinking water didn't meet standards and those questioning when he said the water was safe to drink. Clients threatened to cut his funding if he didn't do what they wanted, even if it would compromise public safety. Edwards ignored them and proceeded as he deemed proper, paying for the work himself. His courage in the face of law suits and threats provide a worthy example for all engineers to emulate. Edward's professionalism was ultimately recognized by the winning of numerous awards and honors, including a MacArthur "Genius" fellowship.

Technical & Project Advisory Roles:

I assisted in the start-up of a program management assignment of an Interstate highway reconstruction in Maryland where Parsons Brinckerhoff was a member of a three-party JV. Parsons Brinckerhoff's local Area Manager, Jerry Jannetti, felt the team could be in trouble without counsel on program management issues and asked me to provide it. The Project Manager was from one of our JV partners, and while normally I didn't give competitors tips on how to run projects better, help was warranted in this case. I advised the project team of several potential program management pitfalls, many of which had never occurred to them. Subsequently, I participated in a workshop with the client to suggest ways the client could get the most out of the program management approach.

In the 1990s, many state transportation agencies hadn't embraced design-build as a project delivery option, but were curious about its potential. I co-hosted a workshop with Tom Warne (former Executive Director of Utah DOT) for the State of Washington DOT on program management and design-build. In that workshop, we used a gap analysis approach to help Washington DOT select their options going forward. The gap analysis established where the DOT was today, where they want to be in the future, the actions to get there, the cost and schedule for each action, the preferred actions, and who had responsibility for each action. For the Minnesota DOT, I organized a telephone conference to discuss design-build and other delivery systems options. My position with regard to design-build is that it is a worthwhile option on most projects. However, one must be aware that it works against design-build as the delivery option when the owner's staff has strong opinions on a project's details and won't allow the design-build firm flexibility to choose among the many methods and particulars.

I provided general advice to our Project Manager (call him Nathan) working on a project's start-up in Oklahoma. The scope required that the firm prepare a geotechnical report for bidders. A geotechnical report would enable contractors to reduce their bids as the report provided the basis for changed condition claims. I said a geotechnical report may lower the contractor's risk, but had the potential to increase our firm's risk. I reasoned that the contractor would claim for more money from the client whenever the actual geotechnical conditions were worse than what was described in the report. If so, I expected the client would go after us for the amount of the claim alleging the firm made a mistake preparing the report. To mitigate the risk, I said Nathan should tell the client to maintain a contingent sum of money available to pay for contractor claims so the client wouldn't be surprised if there were such claims. Further, Nathan should reiterate that contingency in every cost estimate prepared up to the final estimate, so the client wouldn't forget his advice. [*Lesson learned from Paterson's Great Falls project.*]

I was asked by the PIC to help a Project Manager (call him Chet) starting a major transit program management assignment in a western state. Chet was a very senior engineer, who had worked his entire career in the public sector managing major programs, and this was the first project he managed in the private sector. When I arrived at the project office, Chet said, "Why are you here? I've managed projects before and don't need help."

I won him over by explaining I was not there to teach him project management, but to help him understand the nuances of the Parsons Brinckerhoff way of doing things.

Over four days, I reviewed how everything was organized on the project and gave Chet a long list of items to consider; he was very appreciative of my help. The circumstance showed that it's wise to ask for help when one faces a paradigm shift as Chet did — in his case, from managing a project as an employee of an owner to managing one as a consultant.[14-12] Note the parallels to margin erosion Root Cause 2, Unfamiliarity with Client, in that Chet is working for a new employer, something all new hires face.

Chapter 15

INTERNATIONAL EFFORTS

I'd never contemplated working internationally early in my career, because there were so many work opportunities in the US for engineers. Nonetheless, I eventually worked on many international assignments and typically found it a positive experience. Complementing the opportunity to travel and see the world, it was enlightening to discover how non-American engineers accomplish projects and useful to adapt what I learned for domestic projects. This and the following three chapters use case studies to describe efforts related to pursuing and performing international assignments. Chapter 16, Managing a Design-Build Project in Turkey, is especially recommended for someone considering an extended overseas assignment.

Peru:

My first international assignment was in 1977, albeit brief and not successful. I went to Lima, Peru where Parsons Brinckerhoff was teamed with a Peruvian firm pursuing the contract to design a highway project in the Andes Mountains. Before I left for Peru, a principal of the Peruvian firm (call him Jaime) asked me to buy a short-wave radio in the US and said he would pay me for it when I got to Peru. I agreed, and he gave me the make and model number of the radio he wanted.

When I arrived at Peruvian customs, they saw this 2'x2'x2' box stamped with the description of the radio inside and asked if I was importing it or would take it out of the country with me when I left. I sensed if I said I was importing it, I would have to pay a large import duty and was concerned Jaime might not want to reimburse me for the duty. Thus, I said I would take it with me when I left, and the customs agent wrote that fact into my passport. After clearing customs, I told Jaime about the note in my passport, and he said he would take care of it before I left. He then took the radio and paid me for it. Ten days later, I returned to the airport to depart Peru, and, Jaime introduced me to a customs official, who he said was his cousin [*yeah, right*]. After Jaime and his cousin talked off to the side awhile, the cousin scribbled something in my passport, and I had no trouble at departure. To this day, I hope Jaime paid any required import duty. Otherwise, it would be a disappointment to think I may have been used to smuggle electronics into Peru.

My two years of high school Spanish were barely adequate enough to get by in Peru, but fortunately Jaime was more fluent in English than I was in Spanish. While I was there, Lima was under a curfew to control local food riots, and everyone was forced to stay indoors from 10:00 pm to 6:00 am. The riots began after a Peruvian law passed that stated meat could not be sold in the first half of each month. The government wanted people to eat locally caught fish rather than imported Argentinian beef that hurt Peru's balance of payment. I found it unnerving seeing so many armed soldiers patrolling the streets to maintain the peace.

It didn't take long in Peru to realize a few other things. First, working in another country is not the same as being a tourist. While locals generally go out of their way to make tourists feel comfortable, working foreigners have to find their own way to fit into the local business culture. Second, it would be very difficult to find American engineering staff conversant in Spanish who were willing to work for a few years in the Andes Mountains far from civilization. At that time, Parsons Brinckerhoff had only one highway engineer who spoke Spanish fluently. Third, the local firm had little to offer except their alleged contacts with government. While I was struggling to write a comprehensive proposal, they treated the proposal preparation indifferently. Did that mean they were counting on favoritism or bribes to get the work rather than a good proposal? I wasn't sure, but worried payoffs might be involved. In any case, we didn't win the assignment, which probably was the best outcome for us given how difficult it would have been to staff our efforts (or is that just sour grapes?). Although we lost, I personally benefited by the knowledge and experience I gained from working internationally for almost two weeks.

Egypt:

In 1976, Parsons Brinckerhoff started working out of an office in Cairo on a major project for Sadat City, a new city planned in Egypt. Paul Byrne, who would serve as my Deputy Regional Manager seven years later, was the firm's Project Manager based in Cairo. He was returning to the states after three years in Egypt, and I was asked to replace him in 1979. It sounded interesting, and I said I would be willing to go for one year maximum on single-status, as Mary was running a business and couldn't come with me. Our daughter, Stef, was five, and I knew it would be unfair to Mary if I stayed longer. However, the Deputy Head of Parsons Brinckerhoff International, call him Frank, said the Cairo posting was a two-year assignment.

I went home, still wondering what my next project would be. However, a month later, Frank called and told me to pack as I'd be leaving for Egypt in a month. "Okay and it's for one year, correct?" I said.
"No, I told you it had to be for two years," Frank replied.
It took me back a bit, but I reiterated my position, and it was Frank who wound up going to Egypt instead of me. I felt my willingness to go overseas for a year proved my loyalty to the firm and hoped that standing my ground on the duration wouldn't be detrimental to my career.

Hong Kong:

I was asked to manage a highway design project (Route 3) in Hong Kong that would start in 1989. Parsons Brinckerhoff had worked in Hong Kong for a few years, doing mostly mechanical and electrical ("M&E") design for buildings and transit projects. While British civil engineering firms in Hong Kong didn't mind it when we provided M&E services, some were concerned I would be the catalyst for our firm to move aggressively into the civil-highway market. In no uncertain terms, and with curse words thrown in for emphasis, they told Parsons Brinckerhoff's local manager they would pull all M&E work from the firm if I came to Hong Kong. Our local manager reluctantly yielded to that pressure. Thus, my transfer to Hong Kong was cancelled only two weeks before Mary, Stef and I were to leave. We were deeply disappointed we wouldn't be getting a chance to visit and work in the Far East. It was Denver all over again — our lives took a U-turn, and I had to find another project to fill my time. Many Project Managers suffer similar frustrations in their careers.

INTERNATIONAL EFFORTS

Taiwan:

I went to Taipei, Taiwan in December 1989 to lead the firm's proposal efforts for the pursuit of a highway design project, the Nankang-Ilan Expressway. If Parsons Brinckerhoff won the project, Wally Dunn, one of the firm's most experienced Highway Engineers, would be Project Manager. The plan was for me to finish the proposal quickly and return stateside for Christmas and New Year's. However, it became apparent that it would take until mid-January to finish the proposal, and I was asked to stay longer. I said I'd stay if Mary and Stef could join me so we would be together over the holidays, and my request was approved. They came to Taipei and later went with me to Hong Kong when I went there for a short period to help in the production of the final proposal. Everything worked out perfectly, as our team eventually was selected for the assignment.

China (I):

In 1992, I taught a two-week seminar on the design and financing of highways, toll roads, tunnels, and bridges to the staff of the Beijing Municipal Engineering Design and Research Institute. My translator was Ray Choy, an American engineer who worked with me in Hong Kong. It appeared to me as if many mainland Chinese engineers had studied in a two-year technical school, rather than a four-year engineering college. From their questions, they seemed more interested in a cookbook approach to solutions in lieu of using discretionary analyses and engineering judgment. I'm not saying they weren't capable, just that the climate in which they worked in 1992 strongly seemed to favor risk avoidance.

While at the Institute, they asked me to look at two of their active projects. One was a tunnel under the Great Wall of China. There was insufficient parking on the side of the wall nearest Beijing to accommodate all the tourist vehicles coming to the wall, and a tunnel would provide access to a proposed parking facility on the other side of the wall. I commented that perhaps the tunnel should have emergency ventilation in the event of a vehicular fire in the tunnel; I noted that, if a tourist bus caught on fire, the incident would become known worldwide. I said Parsons Brinckerhoff had experts in Hong Kong who could evaluate if ventilation was required, and the firm would do the analysis at cost if the Institute wanted a review. However, they said they accepted the risk and would not determine if ventilation was required, because the national government discouraged hiring international firms as it involved paying in hard currency.

The second project I looked at for the Institute was a bridge over a small river. The bridge was on the highway from the airport into Beijing and was to open in time for the 2008 Olympics. Note the Olympics were 16 years away, but one can never start too early planning for major events.[15-1] The Institute wanted a cable-stayed bridge as a signature structure. After looking at their concept, I said that while the relatively short span length over the river was not economical for a cable-stayed bridge, such a design would look dramatic. I then asked if they had ever designed a cable-stayed bridge before and would need help. They said they hadn't designed such a bridge before, but didn't need help as they would follow the design others in China had used for that bridge type. Using relatively unskilled designers was a high risk in my opinion, but once again, the Institute wanted to avoid paying international specialists in hard currency. It was discouraging that I failed to convince them about the seriousness of potential risks on the two projects, but knew that, at least, I had given them my honest assessment of what they faced.

The seminar's first week was in late November 1991, and I invited several of the Institute's senior managers to an American-style, Thanksgiving dinner at the Beijing Sheraton. The Institute managers were polite, but I could see that American fare was too strange for them to enjoy by the way they picked at the food.

China (II):

Parsons Brinckerhoff was contacted in 1993 to be in a consortium hoping to finance, design, build, own and operate a toll road in mainland China, where Parsons Brinckerhoff would lead the engineering. Our CFO, Rich Schrader, asked me to develop a business plan to assess whether the firm should join the consortium, and if it did, to be resident in China for a year during project start-up. It sounded fascinating, and I agreed and started the process to determine if we should join the consortium.

Early in the process, Schrader said we should give the Parsons Brinckerhoff Finance Committee a head's up about the potential project, as we would need to ask them for funding, if we ever recommended joining the consortium. We got on the Finance Committee agenda to make a presentation on our efforts. While sitting in the boardroom waiting to present, we witnessed a group headed by an employee Director asking for approval to invest in a non-related venture with somewhat similar risks to ours. As they described their proposed venture, it appeared they were recommending it based on gut feelings rather than due diligence, and the venture sounded very speculative to me. I was pleased the outside (non-employee) Directors on the committee tabled acting on the venture until the group had done reasonable due diligence assessing risks and rewards. It showed that the firm's outside Directors were acting appropriately as fiduciary watchdogs of the shareholders' money. To finish the China toll road story, we subsequently identified too many financial risks to make participation advisable and decided not to recommend that the firm pursue the project.

Hungary:

My background in toll roads gave me the opportunity to go to Budapest for two weeks in 1993. My role was as a technical advisor to the team working on the M3/M30 Motorway project in Hungary. Until the collapse of the Warsaw Pact, Hungary had to trade exclusively with the USSR and not with any other country, even the other satellite countries in the Soviet sphere of influence. Thus, all of Hungary's main roads headed in the direction of the USSR, which took a middleman's cut out of all Hungarian goods eventually shipped elsewhere. Now that the Pact ended, Hungary wanted to upgrade its road system to facilitate trade directly with its neighboring countries. Parsons Brinckerhoff won the assignment to study the proposed reconstruction and conversion to toll road of over 50 miles of existing highway from Budapest eastward. During our efforts, I advised our client about both financing and project delivery options.

Canada:

A team of a financial house and an investment bank contacted us in 1996 to provide engineering support services on the Highway 407 project in Toronto, Canada. A few years earlier, the Ontario government spent $1 billion (Cdn) to construct a portion of Highway 407 as a toll road and wanted to get its money back more quickly than by collecting tolls for years. They hoped to sell Highway 407 to a private developer who would build an extension to the existing road and recover its investment by retaining the tolls on the existing road and tolling the extension. I was very familiar with the concept, which is referred to as a public-private partnership, from my work several years earlier in San Diego, which is discussed in subsequent Chapter 19.

Our team won an assignment on the project, and I was the Project Manager for Parsons Brinckerhoff's efforts to develop the engineering concept in more detail and prepare a conceptual schedule. All our US citizen employees required special documentation to enter Canada and work there, even if only for a day.

Once the team completed the assignment, Ontario accepted the concept and sought a team to prepare the final bid documents for the sale of the road to a developer; our team won that assignment also. Parsons Brinckerhoff's role in the second project was to prepare the engineering specifications in the bid documents. When the project went to bid, the high bidder, a Canadian-Spanish Consortium, offered the government $3 billion (Cdn) to buy the franchise. As that amount was three times what Ontario had hoped for, the Province quickly took the money and awarded the franchise in 1998.

Dubai:

Parsons Brinckerhoff was performing program management efforts for the infrastructure portion for construction of Palm Jumeirah on the coast of Dubai, United Arab Emirates. Palm Jumeirah would be a new island, approximately three miles by three miles, in the Persian Gulf in the shape of a palm tree topped by a crescent. In 2002, relatively early in the project, I performed a peer review of our efforts and suggested some modifications to the organizational structure. I also noted several potential risks requiring management strategies including a client new to us and a team created of expats from all over the world. Based on my review, I recommended the Project Manager establish controls to make sure those risks did not impact the quality of our work.

England:

In 2004, I was asked to review transportation project management capabilities in the Godalming office in England, where most of the firm's UK Highway Engineers were based. My review showed the UK and US view job functions and position titles differently and demonstrated there are different ways to structure an organization to complete projects successfully. The review also showed that because most Project Managers tend to be comfortable with an organizational structure they worked with before, they rarely consider if there's a reason to create a different one. A natural reaction, but not always the best way, as projects evolve over time and periodically warrant a new structure to adapt to staff changes or revisions to scope over a project's life.

Panama:

Parsons Brinckerhoff was advising the Panama Canal Authority on its plans to procure a design-build firm to add a third lane of locks. The two existing canal lanes couldn't accommodate the larger ships under contemplation, and the Authority knew it needed a wider canal lane or these larger ships would find other sea routes to deliver their goods.

As planning for the $5.3 billion project evolved, we sensed the Authority felt Parsons Brinckerhoff was not giving the program sufficient senior management oversight. As both a senior manager on major design-build projects and a member of the firm's board, I made several trips to Panama to mitigate those concerns, where I provided technical advice to the Authority and the firm's staff. It always feels good to see the favorable looks on people when I tell them I was involved with this iconic project. After all, this was a prime reason why I became an engineer.

In Panama, we lived in client-furnished housing where we did our own laundry. Some of our team found it burdensome to keep up with their laundry, and thus, the office space we shared with the client occasionally was a bit gamey. When the client complained, the offenders began washing their clothes on a regular basis. I saw humor in the incident and went to the faux-motivational website Despair.com to buy a poster for the office that said:

> **Genius Is 1 Percent Inspiration And 99% Perspiration, Which Is Why Engineers Sometimes Smell Really Bad**

An African Country:

I met with the Ambassador Extraordinary and Plenipotentiary Permanent Representative of an African country's mission to the UN, at Parsons Brinckerhoff's New York office in 2007. The Ambassador said his country was looking for an engineering firm to oversee design and construction firms with whom they were contracting. If Parsons Brinckerhoff agreed to do the work, the country was prepared to give the firm the assignment sole-source. I spoke with the firm's international managers, and they decided to pass on this opportunity. The ability to staff such an assignment properly was too questionable in their opinion. Based on their review of the Transparency International website, it appeared corruption was widespread in the country, and they assumed the country wanted an oversight firm they could trust.

The situation reminded me of an Asian country that was known as being fairly corrupt, where we were advising the government on a new rail project. Another firm was inspecting the rail construction and advised the government that the contractor wasn't complying with the plans. The government told the contractor of the inspectors' concerns. Rather than correcting its defects, the contractor had thugs beat up the inspectors, and the government turned a blind eye to the assault. An extreme example of why it's wise to think twice before deciding to work in a corrupt environment.

CHAPTER 16

MANAGING A DESIGN-BUILD PROJECT IN TURKEY
(With a Contractor as the Client)

While I enjoyed working with Lammie as his assistant, I wanted something more interesting than serving in a staff position. The opportunity to be a Project Manager again came indirectly when I was asked to review our efforts on the 140-mile, Ankara-Gerede Motorway design-build mega-project in Turkey in 1987. The project was a portion of the proposed Trans-European Motorway connecting Gdansk, Poland with Iraq and Iran and was to be constructed using American standards. Our client was two contractors in joint venture (call it CCJV). CCJV had retained our firm to prepare the design and was dissatisfied with our efforts to date. My assignment was to assess the project status and our performance and to recommend improvements.

Based on my review, I determined our Project Manager (call him Mehmet) had lost credibility with CCJV. Regardless of his future performance, I believed CCJV would give Parsons Brinckerhoff a hard time as long as Mehmet remained as Project Manager, and I recommended we replace him. I felt badly about the situation, as I had suggested Mehmet for the position, in part because he spoke Turkish, having been born in Turkey before becoming an American citizen. Working against Mehmet was that some key Turks employed by CCJV knew him from their school days together in Turkey and were envious he was paid as an American.

I returned stateside and began searching for Mehmet's replacement. After a few months with no success, Woody Hitchcock, head of Parsons Brinckerhoff International, asked if I was interested in filling the slot. I was starting to think that way myself and said I'd talk with my family. By then, I had made a second trip to Ankara. Had I not made those two trips to Turkey, I might not have been willing to consider going. However, the trips showed me that the project was exciting and challenging and would give me the opportunity to be a Project Manager again. Another plus was I saw that Turkey would be a remarkably interesting place to work. Hitchcock said anyone willing to work overseas for several years was likely to become a life-long expat. While I didn't think I would want to be an international lifer, I realized how much living overseas, even only a year or two, could be culturally enriching. After speaking with Mary, we agreed I would go to Turkey as Project Manager.

Because I'd be working overseas for our international company, I couldn't contribute to a 401k, and social security taxes wouldn't be withheld. Also, I felt at a great disadvantage when asked to sign a foreign assignment agreement, because I had no idea what pay or perks (living allowance, number of trips home per year, etc.) to ask for. I knew two other American senior engineers were managing mega-projects for the firm (in Hong Kong and Taiwan) and felt I was a better manager than both, with the exception they now had international experience I didn't. Assuming both negotiated fair agreements, I stated I wanted "most favored nation" status. Namely, my pay and perks had to be at least as good as what the other international Project Managers were receiving. Hitchcock agreed, and I signed the agreement committing to stay on the project at least one year. Mehmet then returned to the US, and I took over his company apartment and car and became the Project Manager.

Mary had antique businesses in New Jersey and New York and couldn't accompany me, but our 14-year old daughter, Stef, wanted to come and stay with me, and we agreed she could. We worked out a schedule whereby Mary would visit with Stef and me in Europe or Turkey during Stef's school breaks and my vacation periods. With that plan in place, I was satisfied I could focus appropriately on my project assignment, and my family would have the opportunity to explore new places. At the same time, I realized that for all the cultural benefits for my family and me and for the interesting project assignment, there could be significant stress on the family unit. We hoped some stress would be offset by the meaningful salary increase and additional vacation time I would receive for working overseas. All in all, my family and I entered this new phase of our lives with a fair bit of optimism it would be a positive experience. On reflection 30 years later, I would say the positives clearly outweighed the negatives.

The three years managing the region and the year as Lammie's assistant helped me learn and enhance skills related to leadership, business development, and people management that became valuable when I returned to project management. In addition, I had an understanding of how project efforts function within a consulting firm's organization, as well as a better appreciation of potential margin erosion causes.

I knew the project wouldn't be easy. Nevertheless, I didn't appreciate how demanding it would be to satisfy my desire to return to project management, as the Ankara-Gerede Motorway project proved a very difficult assignment. While it was the most interesting project I ever managed, it also was the most challenging of my career because of all the "firsts" for me. The Ankara-Gerede Motorway was my first design-build project, my first project with a contractor as client, my first project in the metric system, my first assignment living overseas, my first assignment managing a multi-national staff, and my first time as a single parent.

At the time, I hadn't considered that too many firsts were likely to increase the risk for missteps on my part. Was it hubris that I failed to remember that never having managed a project of this type before was margin erosion Root Cause 4? And if the many personal firsts weren't enough, I soon found out the contract fee for the project was wholly inadequate for the task at hand (Root Cause 1). While I was able to overcome all my firsts adequately enough to maintain appropriate quality, I could never develop sufficient budget reductions to make the project profitable. Both the payroll for 140 staff and living allowances for 36 expats in the peak years were too much for me to find ways to generate significant cost savings.

When Parsons Brinckerhoff priced the proposal, the effort involved designing improvements to an existing highway. We expected there would be considerable survey and boring information available along the existing alignment. However, just as we were preparing to sign the contract and commence work, the owner changed the alignment which meant the design now would be for a completely different highway. Rather than widening an existing highway, the project became a new road that would pass through unimproved lands in mountainous terrain. Among other additional efforts, the new alignment required obtaining significant amounts of costly boring and survey data. For example, instead of the originally contemplated five boring machines, we now required twelve machines to meet the schedule, which necessitated importing two into the country. Our costs further increased as geotechnical and survey crews had to camp out in tents for weeks at a time because the new alignment was hours away from paved roads. Aerial surveys were not possible because of the Turkish military's security concerns. These significant additional expenditures meant that upon notice-to-proceed (a year before I began working on the project), the firm was headed for a sizeable financial loss.

As soon as I started on the project, the PIC (call him Jack) and Mehmet told me I had to get CCJV to agree to pay for scope changes related to the alignment shift. I assembled our claim, describing the scope changes from the proposal and the costs for the additional work, and made a formal presentation to CCJV's Superintendent, call him Adam. After hearing me out, Adam, said he was confused and asked why I was making this request. He said the day before signing the contract, Jack and Mehmet were shown the alignment revision and agreed to the change for no additional compensation. Here I was three months into the project, and this was the first I had heard our management had agreed to the change. I'm sure the look on my face showed I had no reply to Adam's statement. If true, it meant any hope of a profit on the project disappeared.

When I had the chance to ask Jack about what happened, he confirmed Adam's version, although he never gave a satisfactory answer as to why he hadn't said anything before to me about the situation. Maybe he had hoped that, as long as I thought it was possible for the project to make a profit, I would find a way to make it happen.

Apparently, Jack and Mehmet, as a Mechanical Engineer and a Bridge Engineer respectively, hadn't appreciated the extent of extra effort and costs required when they agreed to the highway alignment change. Or perhaps, because some staff had already relocated to Ankara and many others were poised to start work, Jack and Mehmet felt they couldn't afford any more staff on idle time before starting work and deluded themselves that the change wasn't significant. In any case, the lesson here is blatantly obvious: Don't agree to something outside your expertise.[16-1]

With an inadequate budget and no additional funding, I had to find ways to cut costs to minimize the substantial, impending loss. I directed staff to standardize design features to facilitate speedier designs. For example, we had to design dozens of huge box culverts under the highway to convey water from one side of the highway to the other; many culverts were twin 20 ft. x 20 ft. boxes over 1000 feet long[20]. Our engineers intended to design each culvert element individually, including designing every slab, wall and roof for the exact height of fill over it. Individually designing culvert elements was too labor intensive for me. Instead, I directed we design culverts as a multiple of 40-foot long standard increments based on only five embankment fill heights: 30, 60, 90, 120 and 150 feet. The highest embankment height over any 40-foot segment set the design criteria for that segment (e.g., if the embankment height over a segment ranged from 30 to 50 feet, it was designed for 60 feet). Thus, we only had to prepare five standard designs to set all twin 20'x20' culverts. We repeated that approach for culverts of different sizes, with meaningful savings in staffing needs, design costs and design time. CCJV approved our approach as commencing construction quickly was important to them.

CCJV had originally bid for a major bridge project over the Bosporus in Istanbul with significant quantities of concrete and not much earthwork. As unit prices typically vary in inverse proportion to the quantities of items, CCJV's proposed prices were low for concrete and high for earthwork. CCJV didn't win the bridge project, but as consolation, was awarded the Ankara-Gerede Motorway project. This highway project would have the same unit prices as in CCJV's bridge bid package, despite the fact the highway had a much higher percentage of earthworks and lower percentage of concrete compared with the bridge project. In other words, CCJV's earthwork unit prices stayed high even with significantly increased earthwork quantities, while structural prices stayed low even though concrete quantities were significantly reduced. Preferring more profitable earthwork rather than less profitable concrete work, CCJV strongly pressured us not to design bridges or tunnels, but to design the road for fills as much as 130 feet high and cuts as much as 130 feet deep.

---

[20] Dimensions are stated in Imperial units in this book even when the work was performed in metric units.

Such pressure was but one way our ethical obligations to serve both our client and the public were constantly tested. We had to determine how to help our client be cost effective without wasting public money and when to contest providing such help. As an example, one part of the project involved excavating a streambed. Our geotechnical engineers added a note on the plans that gravel removed from the streambed was poor quality and shouldn't be reused as gravel elsewhere on the project. Nevertheless, CCJV excavated the streambed gravel and placed it where approved gravel was called for on the plans. The owner's inspector told them to stop and remove any such gravel they placed. At that point, CCJV came to me and said we should give them a letter saying it's okay to reuse the gravel. I stood firm and told CCJV our original finding not to reuse the gravel still applied, knowing it looked to CCJV that I wasn't a team player.

CCJV frequently disparaged any engineer's recommendations, even those of their own engineers, as they always felt a constructor's opinion was primary — and sometimes they were right. A major lesson I learned on this, my first design-build project, was how often designers fail to consider if their design complicates construction. Such lack of consideration on the Ankara-Gerede Motorway project led CCJV to reject some designs and require us to redo them even though the designs satisfied published criteria. For example, we designed the alignment at some bridge crossings on curves or skews. CCJV rejected those alignment designs and directed us to make all bridge crossings at right angles, with no horizontal or vertical curves on the bridge. This approach complicated our geometric design while simplifying our structural design, but more important there were significant construction cost savings to CCJV. A similar issue arose in drainage design, where we determined cross drains were required at 380-foot intervals, and our designers rounded that to 350 feet. CCJV said the reduced spacing increased, by about 10%, the number of times they had to stop earthwork construction to install cross drains (a costly operation) over the 140 miles of highway, and they directed us to redo the design using 380-foot spacing.

I never had to become licensed to practice engineering in Turkey as Turkey didn't have an engineering registration requirement similar to the ones many other countries and the states in the United States have. I did become a member of the Society of American Military Engineers' Anatolian Post, the local arm of SAME in Turkey. That way I could meet with engineers from other organizations in Turkey and stay reasonably current in the profession. I even presented a paper on the Ankara-Gerede project at the February 1988 post meeting.

Parsons Brinckerhoff's stateside offices were seven to ten hours behind Ankara, which meant it was difficult to have questions answered quickly. To save a day in that pre-cell phone, pre-internet environment, I would return to the office after dinner, so I could fax back and forth with the US on key issues in a real-time basis. International phone call rates from Turkey were very expensive, while rates from the US to Turkey were more reasonable. Whenever I wanted to talk about an issue by phone, I faxed a time for the US to call me. Once a week, Mary would call Stef and me at 11:00 pm her time, where the call would wake us at 6:00 am Ankara time. Before reading our notes to all expats describing how expensive calls were, a new expat hire called the US several times in the first week to tell his family and fiancé how he was doing in Turkey. He was stunned at the cost, but out of compassion, we split the bill with him.

CCJV continually criticized our staff and ultimately forced me to replace my highway, bridge, and geotechnical lead engineers. I never was certain if they wanted replacements to improve the quality of the leads or as a power play to show they could replace anyone they wanted anytime. I fought for each lead, but it was always a losing battle.

One person CCJV didn't ask me to replace was my scheduling lead, call him Sean. Instead, six months after I arrived in Turkey, Parsons Brinckerhoff asked if they could transfer Sean to a project in Taiwan. As I thought about it, I realized we could save some money without hurting performance by releasing and not replacing Sean, as his assistant was capable of preparing the schedules going forward. However, I knew CCJV lived by schedules and would fight me on principle if I tried to remove Sean, even if his assistant was qualified. Once Sean agreed to go, I came up with an approach to release him I was confident would work. My plan was based on the fact that Adam frequently complained about some anomaly in one of the numerous schedules, and it was just a matter of time before I would be called on the carpet about schedules. Sure enough, two weeks later, Adam called me in to complain about a schedule issue. The moment he did, I leapt up and said, "I'll take care of this, and you'll never have to deal with Sean again. He's disappointed both of us for the last time, and he's off the project effective today."

Knowing Adam liked decisive action (e.g., getting rid of poor performers), I quickly walked out before he had a chance to ask any questions. Victories were rare on that project, and I still savor that one.

It wasn't easy to recruit or transfer staff from stateside to Turkey. While I was sympathetic to the difficulty involved in getting an American engineer to agree to come to Turkey for more than a few months, I was frustrated when some Parsons Brinckerhoff managers of technical resources in the US avoided my faxes and follow-up phone calls on the subject of staff needs. I overcame that avoidance by calling them at home at 5:00 am their time, as they always answered those calls. At least that way, I could find out if there was any hope they would be sending me someone in the next week or so. All too often I learned I had no choice but to follow Thomas Jefferson's Revolutionary War dictum that, "Remote from all other aid, we are obliged to invent and to execute; to find within ourselves and not to lean on others."

The peak staff of 140 consisted of about twelve expats each of Americans, Brits and Aussies, and over 100 Turks. Each nationality had great pride in their own ability, and, as you can imagine, many felt engineers from their nation were superior to those from the other countries. When I first arrived, there was frequent sniping at each other such as saying Turks are no good at this or Brits are no good at that. While no doubt part of the sniping was related to cultural biases, some possibly related to pay differentials. Americans made 10% more than equally qualified Brits, who made 10% more than Aussies, who received 20% more than Turks. Half our expat staff were supposed to be Americans, because the American Export-Import Bank[21] had money in the project. A shortfall in our

---

[21] The Export-Import Bank of the United States (colloquially called Ex-Im Bank) is an independent, self-sustaining agency supporting U.S. jobs by financing the export of American goods and services.

American staff, had to be covered by additional Americans employed by CCJV, which they didn't want to do. As Americans had so many work opportunities in the US, we had to overpay them to get them to agree to work in Turkey. Thus, I had to utilize some expensive Americans rather than equally qualified, but more economical, Brits, Aussies and Turks which worked against attempts to reduce labor costs.

An international project involves people thrown together for various reasons. In general, Brits and Aussies were quite willing to work internationally, as that was a common practice for their engineers for decades, and their respective tax structures made it favorable for them to do so. Another reason people became expats was related to business expedience — many firms accepted volunteers willing to relocate, regardless of their qualifications. Some expats came to Turkey because they enjoyed working in a remote posting outside their home country. Other times, the willingness to be an expat was to escape problems back home such as family or workplace issues. Thus, there were several people-problems on the Ankara-Gerede team similar to any other organization thrown together based on expediency and individual desires rather than primarily the staff's qualifications and interpersonal skills. It was obvious to me that our team was not functioning well, when I started there. Positions were staked out, and even though the existing approach was dysfunctional, people had settled into daily routines. My task was to quickly eliminate the dysfunction so the team would start working productively with minimal bickering.

Each of the four countries had their major cultural differences. Now, I realize the following are broad generalizations, as there were many who acted differently from others from their country, but here are my thoughts about some differences, anyway. Turks were the most hierarchical of the four, and some called me Bruce Bey. At first, I thought they were mocking me, until I learned that usage was highly respectful and similar to calling me Mr. Podwal. Aussies were the most relaxed, coming to work in shorts and flip-flops in warm weather. They partied hard and worked hard; they called me Bruce as though they knew me all my life. Brits were relatively reserved, wore ties and called me Mr. Podwal, at least at first. British civil engineers tended to have a broader knowledge of both highway and bridge design, than an American who typically focused on only one of those disciplines. Americans tended to be slightly more formal than Aussies, but less than Brits, and acted the most transient of the four, as most viewed themselves as international short-timers.

With those understandings about workers from the four countries, I realized I had to work with the four groups appreciating, not disparaging, inherent differences. To do otherwise, would compromise the ability to get the necessary collaboration from everyone. I put four small flags on my desk (one for each country) and made sure everyone understood I represented the whole team, not just the Americans. I told everyone they could criticize an individual's actions, but not blame a whole nation. At that point, everyone comprehended I wouldn't tolerate nationalistic backbiting, and that I respected everyone's contribution. Things calmed down, and a cohesive team was formed going forward.

One day, the Australian Ambassador to Turkey, Don Witheford, called, asking to visit our office. Of course, I said he could come, but inwardly wondered why he wanted to do so. When he arrived, he said we employed more Australians than any other business in Turkey, and he wanted to find out more about what we were doing. We talked awhile over tea, found out our daughters were schoolmates at the American high school in Ankara, and then he met with the Australians in the office. The visit went very well, and a few months later, Ambassador Witheford invited me to a reception at the embassy. During my posting in Turkey, I also was invited to two receptions at the residence of Robert Strausz-Hupe, the American Ambassador. It was a privilege to represent my company.

When I joined Parsons Brinckerhoff in 1961, the firm had one Geotechnical Engineer in the firm and often subcontracted geotechnical work to specialist firms. By 1987, the firm's geotechnical capabilities had grown tremendously, and I valued all the help it gave me in Turkey. The project, however, required a wider range of geotechnical skills than I previously ever encountered. For example, I needed both a rock consultant who could tell us what to do in 130-foot deep rock cuts, and a soft earth consultant who could tell us how to design 130-foot high fills. When those two specialists arrived in Turkey, I had each give a half-day presentation on their area of expertise, not only to our local 30-person geotechnical team, but also to the Turkish highway agency geotechnical staff. It was a perfect example of providing technology transfer to a client.

At peak, we completed construction plans for four bridges a week and three miles of highway every three weeks. Still, CCJV was always demanding we do things faster, and I feared that eventually a design defect would slip through in our haste and not be discovered until construction. I wanted a formal quality control system to make sure we weren't substituting speed for quality. I'd learned working for Jenny and Lammie that quality doesn't happen automatically and that every interface increased the risk of a quality defect. Thus, I instituted a policy that a set of plans (say highway plans) would not go to construction unless the other disciplines (in this case, the interfaces would be bridge, drainage and geotechnical) had reviewed the plans and signed off on the design. The first time someone came and said there's no time to get all the discipline reviews done or we'll miss a due date, I said we'll do the reviews, and I'll tell CCJV we need a few days to make sure the quality is appropriate. I took the expected tongue-lashing from CCJV, but at least my team learned how important quality was to me. Thereafter, we programmed sufficient time for the quality reviews when scheduling deliverables.[16-2]

Over the years, I've found many face the same question when under time pressure: Do I assume the quality is good enough and make the due date, or do I miss the date while confirming the quality is appropriate? My answer is to check the quality first so you avoid submitting a poor-quality product. Most clients will cut you some slack if you say the work is essentially complete, but you're going to be a bit late while confirming the quality. If the delivery date is so super-critical that missing it isn't an option, submit the work product, but then finish checking it as soon as possible. You want to find any errors BEFORE the client, the bidder or the contractor in the field does. The medicine for correcting an error always is less painful when you find the mistake first and early.

In late autumn 1988, we began developing the conceptual horizontal and vertical geometry for the alignment where the Ankara-Gerede Motorway would cross a small mountain crest. Our conceptual design showed the road would be in 130 feet of cut. We had just commenced taking the necessary borings at the crest to set the geometry and side slopes and do the final design. Meanwhile, CCJV was trying to identify work it could perform during the winter to generate positive cash flow. CCJV said the mountain crest apparently was good rock, and even if it snowed, they could excavate rock all winter. Accordingly, CCJV asked if we could give them the final horizontal and vertical location for the highway centerline at the crest more quickly. I said if we ceased taking borings at the crest, it would be four months before we could finalize the alignment. I then inquired how long it would take them to excavate half the 130 feet. When they said, four months, I said, "I have an idea. Next week, I'll give you a tentative horizontal and vertical fix, plus or minus six feet, and you can start construction. In four months, when you're half-way (65 feet) down, I'll give you the final location, and you can realign the remaining excavation accordingly."

CCJV liked my suggestion and started rock excavation with the tentative fix.

Knowing we'd save money by eliminating borings left me optimistic all would work out well, especially as CCJV didn't ask for a credit for the reduced number of borings we'd be taking. However, soon after starting the rock excavation, CCJV found the rock was so fractured they couldn't do as much winter excavation as they wanted. Further, poor rock meant side slopes CCJV had just constructed with a steep slope had to be flattened, which was a costly retrofit. I shouldn't have been surprised, but CCJV blamed us because the rock was poor quality, ignoring that they never gave us a chance to finish even the first boring.

I understood CCJV wanted to keep its staff and equipment busy all winter, and felt for them that the poor rock increased their costs. It just disappointed me Adam went to his boss and blamed us for the result. In hindsight, I should have written a letter to CCJV describing the risk they faced by not having borings before they started construction. As Jenny always said, "The unwritten documentation letter invariably comes back to haunt you."

Adam never worked on a linear project before, but had been involved with vertical ones, such as power plants. When he saw two used cranes, of a type with which he was familiar, advertised for sale for $250,000, he bought them. Unfortunately, he didn't realize the cranes were more appropriate for constructing buildings than bridges and couldn't lift the large beams typically used on long bridges. When we told Adam the cranes limited the size and weight of bridge beams we could specify, he didn't want to admit to his boss that he made a poor decision and directed us to incorporate the cranes' limitations in our designs. His decision meant many bridge spans were 30% shorter than they normally would be, and thus both costlier and ungainly looking.

Turkey is in a very active earthquake belt. We designed a 260-foot high, multi-span bridge at the northern end of the project to withstand a once-in-250-years earthquake. However, the footings from adjacent piers would be so large they would almost touched each other. Turkish Highway Department told us to reduce costs by designing the bridge to withstand a 200-year quake. Because an earthquake of a greater magnitude would fell all other bridges in the region, they said there's no reason for one bridge to be designed to a higher standard. When I passed that information onto our Structural Engineers, they were opposed claiming it would be unsafe. I told them it's acceptable to change the proposed design criteria, because Turkey has no published earthquake design standard for our guidance, and the Turkish Highway Department understands the risk differential between 200- and 250-year earthquakes and willingly accepts the jeopardy.

My belief is you may follow the owner's wishes when there's no published standard providing the risk can be judged to be within acceptable limits, and you get direction from the owner in writing, indicating the owner understands the risk. After all, any magnitude earthquake will be exceeded someday. Selecting which earthquake return design criteria to use is similar to whether an automobile insurance policy deductible should be $500 or $1000; it depends on how much risk the owner is prepared to accept that something could happen.

One day while walking the proposed alignment, I saw there was a cave in a nearby hill, and someone said it's the site of an old Roman fort. I thought it would be great to add a scenic vista parking area, and maybe even off- and on-ramps, so travelers could visit the fort. However, I was told there are so many Roman forts in the area that one more isn't of interest to anyone. My suggestion went nowhere, although at least I felt I tried to honor a site of possible historic significance.

CCJV constantly directed us to do out of scope work for no additional fee. The contract required us to continue working while claiming and negotiating extras, and we would have to prove breach of contract on CCJV's part to stop working. Our situation was exacerbated by the many expats we had in Turkey whom we would have to continue paying during a work stoppage. Further, we felt it likely CCJV would prevail if we ever sued them, as its ties to Turkey's justice system were stronger than ours.

Accordingly, we continued working while codifying many extras into a $10 million claim. CCJV never agreed they owed us this money, but was willing to pass our $10 million claim to the owner. However, CCJV packaged our claim with a $100 million claim of its own. After CCJV met with the owner, Adam came back and said the owner agreed to 30% of the claims, so we get $3 million and CCJV gets $30 million. As there was little reason why our claim and CCJV's were lumped together into a global settlement, it left a bitter taste in our mouths as we wondered if CCJV negotiated away portions of our claim so they could get more of its own. We doubted our entire $10 million claim would be approved, but leaving $7 million on the table guaranteed we would lose money on the project.

Similar to Westway, the Ankara-Gerede project was one where I both served as Project Manager and supervised the project's staff. A senior deputy, Doug Blenkey, oversaw the administrative tasks so I could focus on other project issues. Blenkey had worked internationally many times and did a great job handling financial, contractual, and expat issues.

As all expats were away from their home countries, I also served as the father-figure concerned for everyone's welfare even outside the workplace. During one cold spell, a British expat came to me very upset because his wife and three young children were almost freezing, as their landlord wasn't providing enough heat in their apartment. I said, "Here's my apartment key. Take the space heater in the living room; it's now yours."

Helping heat an apartment was simple, but keeping expats out of Turkish jails was another story. A British engineer was going through a rough patch and, while in the office, he blurted out, "I can't wait to get out of this [*expletive*] country."
Because cursing the Turkish nation is a crime, the Turkish staff immediately called a meeting to see if they should report him to the authorities and fortunately decided not to do so. Nevertheless, I called the engineer into my office and said, "Go home and pack your belongings. We'll pack your office goods and get you on the first plane out of here."
"But why should I leave since the staff said they wouldn't report me?," he asked.
I replied, "We can't afford to take a chance the Turkish staff will be as forgiving should you slip in the future and say something inappropriate."
I felt we couldn't risk the engineer cursing Turkey again, and the staff reporting him to the police. To me, it was imperative to err on the side of caution in these situations.

There were two incidents where staff spent a few hours in police lockup before we got them released. One incident related to an American engineer hitting a pedestrian while driving on a local street, which can be a criminal offence in Turkey. Luckily, it stayed a civil case as the pedestrian's injuries were minor, and our engineer paid the medical fees, which in Ankara were a nominal cost. The other case was much more serious and involved an Australian engineer inadvertently associating with possible terrorists living next door. A Turkish affiliate arranged for a lawyer to represent our engineer, and, fortunately, the judge released the engineer with his passport while police still were investigating the situation. The lawyer never told me why the judge took the action he did, though I wouldn't have been surprised if money as a form of bail was involved. Before the police completed their investigation, our engineer hurriedly flew back to Australia to avoid any further problems. Dealing with these types of issues wasn't anything taught in engineering school.

When I moved to Turkey, I was a true expat neophyte. Fortuitously, generating a support system was second nature for the British and Aussie expats on our team and their spouses who had worked outside their home country before. The Ankara support system included such things as lists of items to bring to Turkey; the names of American- and European-educated, Turkish doctors; and where to shop locally. Also, the Turkish administrative staff was great and went out of their way helping me settle in. It didn't take long to learn that expats had to make friends quickly, because few rarely stayed in one place for long. You made the effort to seek out others who liked to do the same things as you, to get the most benefit out of living away from home for a brief time.[16-3] Stef is still in touch with students who went to the US Department of Defense's George C. Marshall High School in Ankara 30 years ago with her.

I liked Turkish food from the start (especially the bread, mezze, lentil soup, Iskender kebab and rice pudding), so never went hungry. However, to get my occasional fix of American food, I would go the U.S. military's American Officers' Club, which permitted the non-Turkish, business community to join. I didn't have a satellite dish in my apartment and went to the club to see the weekly NFL football game and even watched the Super Bowl where the game started at 3:00 am in Turkey. I never thought I looked good with a mustache the few times I wore one in my twenties, but decided to grow one to fit in with those in Turkey with a mustache. I doubt I looked more attractive with one, but kept it for my entire stay.

The Turkish people were wonderful; their culture obliges them to help strangers until satisfied they're safe and secure. Once, several of us were walking in a field trying to determine the best alignment for the highway, and we saw a small group of houses in the distance. Before long, a few people from the homes, including the mayor, walked over carrying fruit for us to eat. We were strangers in their area, and they felt responsible to feed us until they were sure we were okay.

At first, I was predisposed to think a Turk who spoke English had to be a better engineer than one who didn't. I quickly was disabused of that notion, as I met many highly qualified Turkish engineers who spoke no English. I realized it would be useful in my work and the off-hours to acclimate to the Turkish culture and took lessons so I could speak some Turkish and hold a basic conversation. Stef has a great ear for languages and said I spoke Turkish with a Bronx accent. On one occasion, I was in a restaurant that I knew over-salted the food, so I asked for my food to be made without tuz, where "tuz" is Turkish for salt. Unfortunately, I pronounced tuz to rhyme with "fuzz," instead of with "news," the correct pronunciation. The waiter looked at me in bewilderment. I found out afterwards, I'd asked for the food to be made without dust.

Our contract said deliverables were to be in English, but CCJV demanded correspondence also in Turkish. I would carefully write out what I wanted to say in English, fine-tuning every word to make sure I used the perfect adjective or adverb so there would be no misunderstanding my meaning. Then, one of our professional translators would convert the letter into Turkish. After I became slightly proficient in Turkish, I would read the Turkish letter to see how well the translator had captured my intent. That's when I discovered translation watered down the subtlety of my meaning as Turkish has far fewer words than English.

Even after I spoke some Turkish, I brought a translator when I met with someone who didn't speak English. The translators weren't professional translators, but engineers who understood technical terms. I learned to speak no more than a few, short sentences before pausing to have it translated. Anything longer and it was likely these amateur translators would forget something I had said. To be helpful, translators sometimes answered questions without deferring to me for the reply. Once I realized that might happen, I forewarned them not to do so, as it was inappropriate if I, the senior representative, was not the one responding.

One of the firm's American engineers came to Turkey and made no effort to learn even a few words of Turkish. Whenever he wanted something, he asked for it in English. If the person he was speaking to didn't understand him, he spoke louder, and if that didn't get him what he wanted, he shouted it in English. He was very relieved when his assignment ended, and he returned stateside. Everyone else was also.

In addition to trying to speak Turkish, I had to learn new physical cues. Body language differs from country to country, and I learned which hand gestures, table manners and such are proper in Turkey and which are deemed offensive. Before taking a foreign assignment, I recommend reading a book or internet article about communicating by body language in the country you'll be working.

Ankara turned out to be a good place to work. It has far fewer touristic distractions than, say, Istanbul; thus, it was easier for expats to focus on the work that had to be done. Ankara is Turkey's capital with many embassies and has a major engineering university, Middle East Technical University ("METU"), where classes are taught in English. American workers in the office registered with the American Embassy and received frequent letters from the US Department of State advising how to avoid terrorists, such as by varying the route to work. No one I knew paid any attention to these letters, as everyone felt the inconvenience outweighed the risk potential.

A four-story house in a residential area served as our workplace; my personal office was in a former bedroom on the third floor and had a private bathroom. I used the tub to hold supplies and my coffee machine. Nearby were numerous embassies, each with a guardhouse manned by a Turkish soldier carrying an automatic weapon. I was asked if we wanted a guardhouse with an armed guard in front of our building and declined. I'm a realist and felt one guard wouldn't be able to stop anyone wanting to harm us. We did have doormen in the building, whose function was more for courtesy than security. The doormen felt they failed at their job if I opened the front door myself. To avoid embarrassing the doorman if he was away from his post, I made a lot of noise when leaving to give him time to get to the door before me.

Inflation in Turkey was running over 70% annually, and we gave local staff a 40% pay increase every six months so they could afford to buy food. About half our costs were in Turkish Lira, and inflation made it difficult to estimate with accuracy what it would cost in US dollars to complete the project.

Project staff worked very hard, including half-days on Saturdays. A number of them worked on Sundays as well. I came to the office on both Saturdays and Sundays so the staff knew I was aware of their hard work and supported their willingness to put in all those hours without additional pay.[16-4] Turkish staff had so much pride in being part of the team, they were willing to work any days, any hours, asked of them. I kept my eye on anyone who seemed especially stressed, typically expats with family in Turkey, and would tell them to cut back on their hours.

One day I received a letter from CCJV stating they were going on a 13-day work cycle with one day off every other week and demanding we institute a 56-hour workweek to keep up with them. I went to Adam and said I refused, but he said that under the terms of the contract I had to comply with their request. I countered that our personnel were already working an average of 72 hours per week on a voluntary basis — that if I ordered them to work 56 hours, they would slow down. That became the only time Adam backed down on a directive.

The vast majority of Turkish staff were non-devout Muslims. During the month of Ramadan, staff who were devout Muslims couldn't eat or drink during daylight hours, and overall productivity was lower in general. The office print room was in the basement, and print room staff drank milk to relieve discomfort from the ammonia smell. Because most print room staff were devout and couldn't drink milk during Ramadan, we knew in-house printing would be slow that month.

Working in a new environment, meant I had to become familiar with local, design concepts.[16-5] For example, although I had designed local streets before, I couldn't understand why many curbs in Ankara were about 18 inches high. I'd designed 12-inch curbs in the past, but never more than that. I found out that because Ankara lacked a good street drainage system (in the 18 months I was in this city of over three million people, I saw only three catch basins), the high curbs helped channel rain flows. Also, high curbs deterred drivers from driving on the sidewalks. One of my scariest moments was being driven on an Ankara sidewalk at 20 mph in reverse. Pedestrians calmly stepped aside to let the car pass them.

150   THE ENGINEERING IS EASY

In addition to running the Ankara-Gerede project from the Ankara office, I served as Country Manager for Turkey and President of the firm's Turkish subsidiary. Those roles meant I oversaw a small Turkish staff working on projects in an office in Istanbul, an hour's flight time from Ankara. A year into my tenure in Turkey, the Istanbul Office Manager told me he just signed a contract to design a skyscraper. When I checked into their tall building experience, I found the Istanbul structural engineers had designed only small buildings before, never a skyscraper. I became concerned as I thought of all the risks we were blundering into: A new client, a new discipline, and a Project Manager who never managed a similar project. Thankfully, before I had to come up with ways to address the risk, the client canceled the project soon after we started.

The Turkish skyscraper incident came to mind two decades later when I found out the firm was designing an unusual skyscraper in the Middle East whereby each floor of the building revolved slightly from the floor below. The final configuration would have the highest floor rotated 180° from the ground floor. As I didn't know if this would be the first building of that type the structural engineers in that office had designed, I feared an extreme event, such as an earthquake or severe windstorm, could result in loss of the building's cladding or even its structural equilibrium. To mitigate my apprehension, I had Mike Abrahams, Parsons Brinckerhoff's award-winning structural engineer based in New York City, review the design and offer advice if necessary. Abrahams determined the current engineers had the required skills to complete the designs. Better safe than sorry.

Back then, flying in Turkey was different from what I'd experienced elsewhere. In 1987, Turkish Airlines permitted smoking during in-country flights, and three of us flying together to Istanbul asked for seats in the non-smoking section. We discovered the plane had an interesting seating arrangement — every other row was non-smoking, so we had smokers in front of and behind us. I beat the system on our return flight by booking us in a smoking row, which meant we had non-smokers front and back.

Because Mary worried about terror threats against US airlines, she insisted I fly on non-American carriers, and I flew back to the US on Swissair in late December 1988 for some meetings and the holidays. While in the New York Office, I sat at Warren Buser's desk as he was on vacation. Fifteen years earlier, he and I had bowled together in the company league. Buser was a senior Port Engineer who recently completed a long and arduous assignment for Parsons Brinckerhoff as Resident Engineer in Somalia on the Port of Kismayo project. He earned many frequent flyer miles traveling to and from Somalia and was using those miles for a vacation in England with his son and daughter. They were flying back on Pan Am 103 when it exploded over Lockerbie. The shock we all felt brought home the randomness of everything.

A traditional American competitor of Parsons Brinckerhoff opened an office in Izmir, Turkey and was starting to design a highway similar to the Ankara-Gerede Motorway, but for a different design-build contractor. I flew to Izmir to visit the design firm hoping to get work for our staff. I stated there were almost no engineers in private practice who had designed a highway in Turkey, but we have 100 Turks and 40 expats on the Ankara-Gerede project that had been through the learning curve and now knew what to do. I offered to subcontract to the firm saying it would be more cost effective than if they created an organization from scratch, as we had to do. The firm's manager figuratively threw me out of the office, saying they didn't need our help. He had too much pride to accept that giving us a portion of the work would increase the likelihood for profit. Several years later, I spoke with that firm's President, who told me the project turned out to be their biggest money loser ever. History repeated itself as their learning pains in Turkey mimicked ours.

Failing to get any work in Izmir meant there was no chance we would have enough work to keep all our staff busy and employed when the Ankara-Gerede Motorway project ended. We continued marketing other clients, including the U.S. Army Corps of Engineers, but with limited success. Unless staff knows there's future work for them, they either will start to drift away to other firms or will stay, but with little enthusiasm, until work runs out and they are terminated. The best way to keep staff on board and productive to the end of a project is to make it clear that those who stay will receive appropriate severance/bonus packages.

One day, I visited a Turkish Ministry to market them for work and found there weren't any assignments for us. While there, the minister lamented that he had a problem with government-owned land that some feared had been contaminated by radioactivity released from the Chernobyl nuclear power plant in Ukraine two years earlier. No one wanted to build on the land, even though the minister claimed tests showed the land was not contaminated. I said if you know it isn't contaminated, it would be a powerful statement to make if you were willing to take some soil from the site and spread it in your front yard. At that point, he changed the subject, implying either he knew the soil was contaminated or he feared what his wife would do to him if he suggested dumping the soil near their home.

A year and a half after I started on the Ankara-Gerede Motorway project, I was asked to return to the US for my next project. CCJV agreed I could turn the project over to my deputy, Jim Rozek who was one of the hardest workers I ever met. In addition to long hours on the project, Rozek concurrently taught a course on highway engineering at METU.

I had enjoyed working internationally and now was experienced working for a contractor on a design-build project and in metrics with a mostly non-American staff. While my next assignment was in the US, I hadn't shut the door to another international assignment. The opportunity to see more of the world as an expat was a strong pull for me. Nevertheless, I understood that what was good career-wise could have negative impacts on my family, especially when the family is split for an extended period.

For example, it was not easy to deal with personal issues while assigned to Ankara. I was back in the US on business and was at home with Mary one evening in New Jersey when she fell seriously ill. I had to delay my return to Turkey to be with her until she recovered. A simpler issue was when Stef had to decide with whom to celebrate her 15th birthday. I worked it out whereby she spent breakfast with me in Ankara and then flew back to the US to have a birthday dinner with her mother; the seven-hour time difference and some money for airfare made all that possible.

CHAPTER 17

MANAGING A PROJECT IN HONG KONG

Two years after my earlier false start in relocating to Hong Kong, I got a second chance to work there. A JV of Maunsell (a British firm I worked with in Turkey) and Parsons Brinckerhoff was shortlisted to be interviewed for a Hong Kong highway/tunnel project, the preliminary design and environmental assessment of the proposed Central Kowloon Route ("CKR").

The interview took place in 1991, while Hong Kong was a Crown Colony of Great Britain, and six years before it was turned back to mainland China. Parsons Brinckerhoff still was hoping to gain a stronger foothold in highway design in Hong Kong, and I was asked to be the prospective JV Project Manager. Maunsell, in lieu of blackballing Parsons Brinckerhoff from anything related to highway design, teamed with us because they wanted our M&E experience for the tunnel design and knew they needed someone with my skills as Project Manager to win the assignment.

It was the first time the Hong Kong Highway Department used an interview to select a firm for a project. When our team assembled for interview rehearsals, I noted British and Hong Kong engineers are much more reserved when pursuing work than Americans are. In the US, we make a major effort to demonstrate why we're the best, while in Hong Kong it was poor form to brag about your credentials. Even though I decided to ratchet my enthusiasm down by 50 percent, I still was the most animated of our team at the interview. Regardless, what we did was successful, and our JV was chosen for the assignment.

I would be on single status for this overseas assignment, as neither Mary, who still had her business to run, nor Stef, who was in college, could relocate to Hong Kong. Both came out for short periods and stayed with me in a flat in Stanley on Hong Kong Island, which made the 15 months I was away from home more tolerable. The foreign assignment also enabled us to visit many new places: Stef visited Singapore, Mary and I toured Thailand, Australia and Japan, and the three of us got to Beijing.

As in Turkey, work in Hong Kong was in metric units; however, there were important differences in those two international assignments which are noted throughout this chapter. These differences emphasize that the more one knows in advance about how things are done in an international setting, the more likely both one personally benefits from the posting and their work products are successful. Thus, before agreeing to a foreign assignment, one should try to learn what to expect by speaking with as many people who have worked in that setting as possible.

In 1991, the Parsons Brinckerhoff Hong Kong office was the firm's largest office with over 400 staff. However, when I arrived in Hong Kong, I found the office lacked a support group to help expats as there had been in Turkey. It also seemed many Hong Kong admin staff viewed expats (me included) as a burden to avoid. Here I was trying to start a new project, while simultaneously looking to find a place to live and then furnishing it, all in a location that was foreign to me. It was extraordinarily frustrating.

I had worked with one of the Hong Kong expats, Christian Ingerslev, a few years earlier in New York, and he and his wife, Harriet, went out of their way to help me settle into the routine in Hong Kong. They were kind enough to have me over for dinner a few times and even took me sailing at the Hong Kong Yacht Club. Mary was in Hong Kong on Election Day in 1992 for the U.S. President, and she and Harriet decorated the ballroom where many Americans went to hear the election results. We had to cast absentee ballots that year.

I worked most of the time in Maunsell's Kowloon office where the majority of the CKR design was performed. While there, I often was the only American in a 400-person office of mostly Chinese with some Brits and Aussies. I never sensed I was fully valued, perhaps because Maunsell was one of the British firms that wasn't happy about an American firm expanding into highway design in Hong Kong. It was common in the Maunsell office for people who were celebrating something important (e.g., their marriage or birth of a child), to buy individual pastries for everyone. So, on the Fourth of July, I bought pastries for the entire office, just in case some people didn't remember who won the American

Revolutionary War. I'll accept that maybe I went overboard a bit, but it was fun.

Unlike Turkey, which was based on American standards, the CKR design used British standards and terminology, including designing for right-hand drive. Another difference was our designs and correspondence were all in English, and translation to Chinese weren't necessary. I learned to speak only a few Chinese phrases and carried cards with my home address written in Chinese so taxi drivers would know where to take me. Soon after arriving in Hong Kong, I sat with an expat engineer from the UK and went over a list of British words and terms to help me understand them and communicate effectively with the project team. As examples, I learned a footpath in Hong Kong is a sidewalk, and a lorry is a small truck.

The Hong Kong Highway Department gave us a conceptual design as a starting point for our efforts. Kowloon is a peninsula pointing south, and CKR's alignment would bisect it going from west to east. The highway would serve 150,000 people expected to live and work in a new development on the site of an airport to be replaced by a new airport under construction on Chek Lap Kok Island. The conceptual design had CKR starting in west Kowloon at an interchange with a road along the waterfront. Then, CKR would be on viaduct in the center of a street until entering into a tunnel section beneath a large hill in central Kowloon. Upon exiting the tunnel on the east face of the hill, CKR once again would be on viaduct in the center of a street before connecting with a harbor road along Kowloon's east side. In the future, once the existing airport closed, the highway would be extended further eastward on a filled-in bay towards the new development.

The CKR viaduct sections in the street centers would be positioned between high-rise buildings, for both the east and west approaches to the tunnel through the hill. That concept meant the viaduct parapet would be twelve feet from the third floor of adjacent buildings. To minimize noise impacts, our environmental subconsultant proposed enclosing CKR with a roof, similar to an existing Hong Kong viaduct, and the client concurred with that recommendation. Obviously, an enclosed viaduct results in a more robust and costly structure because of the additional structural loads.

Further, besides the extra structural costs, enclosing the viaduct requires additional lighting and a ventilation system, meaning an enclosed viaduct is relatively expensive.

To me, an enclosed viaduct was a costly, elevated tunnel which can be even more intrusive on the urban landscape than a standard viaduct. Since we already had a tunnel section through the hill with typical ventilation and other tunnel systems, I thought placing all of CKR in an underground tunnel should be an option. I directed we compare the cost of (a) the currently contemplated enclosed viaduct-tunnel-enclosed viaduct concept with that of (b) placing the entire highway in an underground tunnel. It turned out the cost of construction and of land acquisition favoured [*note the British spelling*] option (b), an underground tunnel, over the enclosed viaduct sections. That cost saving plus the visual improvement to the neighborhood clearly made the underground tunnel option the preferred alternative. The client stated the new concept was, "Brilliant," and we earned kudos for suggesting it.

Motorists passing through the CKR tunnel would pay a toll, and we were to design the toll plaza. I asked about the revenue and cost analysis that determined the toll rate to use so toll revenues would be sufficient to pay off the bonds that would finance CKR's construction. "What bonds are you talking about?," was the response, "Hong Kong has $15 billion (USD) on hand and can build this project without bonds. Collecting tolls is only a symbolic action as drivers expect to pay to use a tunnel."
I never before or since had a client with so much available cash. It was rumored the UK was looking to spend all of Hong Kong's money before the Crown Colony reverted to China.

We submitted our first project deliverable a few months after starting the CKR project. I asked when we could expect the client's comments, as I was accustomed to a project hiatus in the US while the client reviewed our work product. I was told it isn't done that way in Hong Kong. It would be disrespectful for the client to make a detailed submittal review, because that would imply it thought the engineering firm it retained didn't know what it was doing. Of course, the client looks at the submittal and comments if it notices something significant; it's just that they don't comprehensively review the submittal. Thus, I was told it was safe to commence the next phase of work immediately after making a deliverable. It was liberating knowing my client trusted me.

In addition to project management duties, I also designed portions of CKR. I did the design the old-fashioned way, without using computer aided design & drafting ("CADD"), which was becoming the worldwide norm. I didn't know it then, but that was the last time I would perform basic design. Subsequent clients for whom I worked used various CADD systems, and becoming proficient at different CADD highway design systems was not the best use of my time and skills. Also, others could do design more economically than I could. Thus, after Hong Kong, I limited my efforts to project management tasks with periodic reviews of designs prepared by others, rather than creating designs myself.

I decided to write a paper on the CKR project for the annual International Bridge, Tunnel and Turnpike Association conference, and the client's Project Manager agreed to co-author it. He reviewed my draft paper, proposed edits and was pleased with the result. My client was happy, and I had another paper to add to my résumé — a clear win-win.[17-1]

In 1991, the United Nations held a two-day meeting in Bangkok, Thailand, to discuss possible alternatives for both a new highway and a new railroad to connect Southeast Asia to Europe. The five permanent members of the Security Council and most Asian countries attended the meeting, and the International Road Federation ("IRF")[22] was asked to provide a delegation. Parsons Brinckerhoff's former CEO Henry Michel was serving a term as IRF President and invited me to lead the IRF delegation that included members from two other consulting firms. The delegation was consulted on several occasions during the meeting for our position on issues. I wrote our comments out beforehand, so simultaneous translators could keep up with me when I spoke. It was fascinating to participate and observe the northern and southern countries jockeying to get support for

---

[22] The IRF is a global not-for-profit organization providing knowledge resources, advocacy services, and continuing education programs.

the highway and railroad alignments to pass through their part of Asia. It was only 17 years since the Vietnam War ended, and I was surprised when the Vietnamese delegate asked to take a photo with me. He and the Outer Mongolian delegate each invited me to visit their country, both of which I was unable to accept. Leading the IRF delegation was a privilege that flowed from agreeing to work overseas.

I'd heard no American Civil Engineer had ever become a Member of the Hong Kong Institute of Engineers ("HKIE"), the licensing organization for Hong Kong engineers. The rumor was British engineers in HKIE discriminated against Americans (I imagine no less so than American engineering licensing boards discriminate against British engineers). I applied for HKIE membership and completed the paperwork; the only thing left was to be interviewed by two HKIE members. The two members stalled setting an interview date until after my Hong Kong assignment ended, and I'd returned to the US. As a result, I never was licensed in Hong Kong, so I guess the rumor was true.

Income taxes are withheld regularly from U.S. paychecks, making reconciliation relatively easy at year's end. When I'd worked in Ankara, Parsons Brinckerhoff paid certain Turkish taxes which meant its expat employees did not have to pay income taxes to Turkey. The Hong Kong Crown Colony government didn't withhold anything for taxes, and I knew I'd have to pay taxes after I finished working there. I had returned stateside, and my accountant said I owed $25,000 (USD) in taxes to Hong Kong for the 15 months I was there. I didn't have that much cash available, so I wrote the Hong Kong Tax Department to say I'd send them $1000 a month for 25 months. They wrote back saying that wasn't acceptable, and I had to pay the entire amount immediately. I responded that the best I could do is increase the monthly payment to $1500 and enclosed the first check. They cashed the check and told me to keep them coming. I didn't return to Hong Kong until I paid off my entire debt. I had learned I had better understand the local tax code before working internationally again or else I might wind up in a similar situation as in Hong Kong.[17-2]

CHAPTER 18

MANAGING A PROGRAM IN GUAM

In 2007, Guam Department of Public Works ("DPW") was seeking a Program Manager for planning and design management of $200 million in construction of roads and bridges. These projects were required to handle traffic from a planned relocation of a U.S. Marine base from Okinawa to Guam. The new base would increase Guam's population by almost 20% and severely strain the island's transportation network. Program management efforts would include development of an extensive public involvement program, design of $40 million of the traffic and road improvements, and technical and administrative management of several consulting firms performing design under contract to the DPW.

Guam, an American territory, is a picturesque Pacific island some eight hours flying time west of Hawaii. A marriage between an international assignment (which I always enjoyed) and a domestic one, the opportunity of working in Guam definitely interested me.

Parsons Brinckerhoff submitted a proposal and was shortlisted to make a presentation. At the presentation, I described how a Program Manager works to show I understood program management comes in many forms and that I was prepared to work with the DPW, however they were comfortable. I demonstrated three different options for accomplishing a task: Sometimes we perform a task on behalf of a client, sometimes we share a task's performance with a client, and sometimes we assist a client who performs most of a task itself.

Our formal presentation included a PowerPoint slide show. As our speakers gave their individual presentations, they would nod at our technician running the computer whenever they wanted him to change the slide on the screen. At one point during my part of the presentation, I was looking at the interview panel while nodding that I wanted the next slide. When I noticed the panel looking quizzically at the screen behind me, I turned to look back at the slide and saw a view of the Manhattan skyline and Brooklyn Bridge. Apparently, the technician accidently hit the wrong key, and his screen saver popped up. I turned back to the panel and said, "And if you hire us, this is what Guam will look like in five years."

It was corny, but it caused the panel to laugh, which was much better than everyone focusing on our miscue.

The interview room was quite large, and we were seated relatively far from the interview panel during the formal presentation. When we finished our presentation and went into the Q&A phase, I had our team move their chairs so they sat much closer to the panel. Guamanians strongly believe in the importance of personal relationships, and our moving closer was deemed a very positive act. In addition, the selection panel wanted an ethical process and may have been influenced further in our favor because we didn't pressure Guam politicians to help us get selected. Overall, what we did resonated with the panel, and we were chosen to negotiate a contract.

With that win, my record was 15 wins out of the last 18 pursuits that went to interview where I would be Project Manager. Over the years, I had several supervisors during these pursuits and was sure few had an appreciation of what I had accomplished. Realizing no one will advocate as well for me as I could, I gave each supervisor a written summary of my won-loss record after each pursuit.

A problem arose during the contract negotiations that lingered over us for months. The Parsons Brinckerhoff planning group had proposed a budget for their portion of the scope, and the client felt the budget was way too much. During fairly intense negotiations, the group's leader (call him Mitch) told the client he would reduce the group's budget significantly. I called for a caucus and pulled Mitch outside the room. When I asked him to define the assumptions he now was making such that he could cut the budget, he couldn't think of a single one. It made me wonder how Mitch developed his initial budget if he could reduce the budget so significantly without modifying scope. Not surprisingly, the client had similar concerns. Rather than saying they were pleased we cut our cost, the client said they felt we had tried to take advantage of them with our initial, excessive budget.

MANAGING A PROGRAM IN GUAM

We looked bad because Mitch apparently hadn't prepared the original planning group budget himself and thus wasn't able to respond to detailed questions. Mitch should have called the actual preparer of the budget and gotten more info before acting hastily. The lesson here is never cut a budget unless you can explain the basis on which you reduced it, such as a reduction in scope or risk.[18-1] As examples, you can say that you reduced the budget because you now understand there will be two fewer alternatives or you now are convinced the existing data is sufficient and won't require collecting supplemental material.

Three lessons I'd learned in Turkey also proved applicable once we commenced working on Guam's program management services contract:

1. We had no permanent staff in Guam before winning this project, and it was difficult to assemble a sufficient number of qualified permanent staff to a remote site. All too often an international Project Manager is forced to beg for support from stateside offices, and timely support still may not come.[18-2]

2. Stays of less than five full days in a remote office are rarely cost effective. It's best to avoid using short-timers who will spend more time traveling to and from the location than working in the office.[18-3]

3. It's easier said than done for a new branch office to get administrative support from headquarters. For example, a new office needs such things as furniture, computers, local bank authorizations, business cards, and telephone services. While there are ways to meet all those needs, a new office's staff are either recent hires who don't know whom to ask for what in the organization or transferees who don't know local practices. In Guam, the office restroom lacked toilet paper and paper towels (until I bought them with my own money), because stateside hadn't yet authorized purchase of supplies by local staff.

There were two unusual situations in Guam we had to consider when performing the work. First, unit prices for construction on Guam are relatively high, because many construction items (e.g., asphalt) had to be imported and there was a shortage of skilled labor on the island. The second related to the fact that Guam's population was about 3000 when the island was acquired by the USA after the Spanish-American War. That small base of people means many of today's Guamanians are related to each other, and you had to be aware there was a potential conflict of interest at every turn. For example, practically everyone we interviewed for a position or the owner of every local firm we met with to discuss an assignment had a relative in a key government position. You also had to avoid talking poorly about anyone, even in jest, because the person you were speaking with could be a close relative of the person you were disparaging.[18-4] Of course, not just on Guam, when you speak negatively about anyone, it should only be in a constructive manner.

Soon after starting work on the project, Guam Governor Felix P. Camacho requested that we meet and then hold a joint press conference. I asked my staff if I should wear a suit and tie to the meeting/press conference. They said the Governor would wear an island shirt, and I should do the same. It seemed too casual to me, but I wore a short-sleeved, dark blue island shirt with palm trees on it. I shouldn't have doubted my staff, as I fit in perfectly with Governor Camacho.

The DPW stated they had never done a project like this before and were counting on us to tell them what to do. However, I never got the chance to put my thoughts into action, as the DPW head, call him Vic, revised our scope almost weekly in reaction to the latest political demands and pressures on him. Unfortunately, it's almost impossible to align scope, schedule and budget on a rush-rush project with frequent scope changes.[18-5] You rarely can satisfy both your own firm's internal business management system and the client's expectations in that scenario.

Normal start-up problems were exacerbated even more when DPW gave us notice-to-proceed on technical tasks, but deferred authorizing us to perform project management oversight tasks. The inability to develop and implement project policies and procedures, including those related to quality, schedule and budget control, significantly hampered our early attempts to begin the project in a logical fashion. It was a disaster waiting to happen — and it did.

As a cost control measure, we decided to have a relatively small staff on Guam and used an arrangement whereby our Guam office was supplemented with seven stateside offices and the offices of five Guamanian subconsultants; this approach is called a virtual office. I knew the more offices involved with a project, the more likely something will go very wrong before it's noticed by project management. To overcome a virtual office's disproportionate number of interface risks, one must budget for and carry out frequent visits to each remote office to assess performance, as managing by teleconference and e-mail works only up to a point. The alternative is to expect to be disappointed.[18-6] Unfortunately, without the client's authorization to work on management tasks, our stateside oversight was not anywhere as rigorous as it should have been, and many of our early delivery dates were missed. A representative problem occurred when a stateside office stopped working on the Guam project to pursue work from its local client without telling anybody on Guam.

Obviously, such things as tracking of deliverables would have been better had DPW approved our starting on project management tasks. Once again, the fault was mine. I should not have accepted a limited NTP, even though several offices were pressing me to start on the project so they would have something for idle staff to do.

Jenny taught me that having minutes of even a casual conversation is indispensable to avoid problems arising as there always will be people forgetting or misinterpreting what was said.[18-7] Therefore, I took notes whenever I'd meet with Vic. One day, as I was taking notes while Vic was speaking, he suddenly stopped talking and directed me to stop writing. He said it disturbed him when I took notes, since everything he was saying he had said before; he wasn't accurate, as I had noted several clarifications to things he previously said. I was taken aback by his directive as having a contemporaneous record is a given to me. After all, poor documentation came back to haunt me both on the New Jersey toll road project where I failed to define the project we thought we were designing and on the Turkish project where I failed to document that the client wouldn't give us time to take borings at the mountain crest.

Vic's instruction hung over me whenever we met thereafter, and I now was forced to write notes as soon as I left meetings by relying on memory. Further, I felt constrained from sending a copy of my notes to Vic to confirm what I thought he said. Without written confirmation, we lacked the ability to justify why we had to revise our efforts whenever he modified a previous instruction. It was very frustrating, as we could only speculate on the reason for Vic's directive. It reminds us that when we are under constant pressure from those above, it can lead to poorly thought out decisions as we struggle to handle the stress. We must take the time necessary to avoid being rushed when we can.

To stay within the client approved budget, the plan was for me to shift off the Guam project after six months and turn the project over to my deputy. That plan was accelerated at Vic's request four months into the project, when he took umbrage after I used the phrase "dog-and-pony-show" when describing a pending presentation. I was using the phrase in its contemporary meaning to be something impressive: an elaborate display or presentation to promote something. However, Vic thought of the phrase in its original meaning as "a small-town, circus-like event of little significance." Because he felt I wasn't taking the upcoming presentation seriously, he directed that I be replaced as the Project Manager.

I was very disappointed the client wanted me relieved, but accepted Vic's request calmly. After all, it only meant leaving Guam two months earlier than planned. Not wanting to leave on a sour note, I went to Vic to say my good-byes and tell of my warm affection for Guam and its people. Although I was no longer Project Manager, I continued to provide advice to the project management team.

Apparently, my willingness to be gracious was duly appreciated. Several months later, Vic publically apologized to me at a project meeting, saying he now regretted asking that I be replaced. By then, Vic had requested replacement of the Project Manager who replaced me and, soon afterward, of our third Project Manager.

CHAPTER 19

MANAGING A PUBLIC-PRIVATE PARTNERSHIP PROJECT IN SAN DIEGO

By 1990, governmental agencies were becoming creative finding ways to finance infrastructure projects other than by increasing taxes or user fees. One financing approach gaining credibility was the Public-Private Partnership ("P3"). In a typical P3 road project, a private developer finances the design and construction of the road and makes money by charging tolls on the completed project. In that scenario, it would be the private developer, and not the governmental agency, who takes the heat from the public for charging tolls.

California passed a law authorizing Caltrans to issue four pilot P3 franchises for either a highway or transit project. A franchise holder, a private entity, would fund obtaining environmental approval, designing, constructing, operating and maintaining the project for 35 years. The public entity, Caltrans, would allow the franchise holder to use eminent domain to acquire property; Caltrans would own the project, assuming full control after 35 years.

California Transportation Ventures ("CTV") was formed to win one of the franchises by four firms: Parsons Brinckerhoff, Transroute (a French firm with toll road operating experience), Fluor (a design-construction firm), and Prudential Bache (a financial house). Each firm committed to 25% of the costs to pursue the franchise and, if successful, to finance, design, construct, operate and maintain the road. CTV planned to recover its investment by collecting tolls or selling the franchise to another party. Parsons Brinckerhoff, Transroute and Fluor each proposed someone to be CTV's President, and I was chosen as President and de facto Program Manager, with Transroute providing my deputy and Fluor the Chairman.

To run CTV for the four partners, I would have to manage all elements of the program, including raising the project's financing if we were granted a franchise. For the first time, I would lead the effort to estimate an enterprise's costs and revenues over four decades using various pro forma analyses. It meant I quickly had to learn how to prepare company financial statements (including balance sheets and earnings & revenue forecasts) and to evaluate funding strategies (such as a potential public offering or bonding option), as those on the CTV Board were already expert in those areas. I read books on the topic of project and corporate financing and talked with several specialists. Fifteen years later, the knowledge I gained understanding company financials and corporate funding options proved invaluable when I was elected to the Parsons Brinckerhoff Board.

On my first day at work in CTV's Orange, California office, there was a 5.6 magnitude earthquake, centered 50 miles away. I started to panic as the building shook for ten seconds. A California native from Prudential Bache was at my desk using my phone when the quake hit and never moved from the seat. He continued talking on the phone saying things such as, "You feel it, too? Yes, it's a big one."

His calmness reassured me. While I initially had thought of racing for the doorframe, I decided it was wise at least to move away from the windows. When the quake ended, separately, four people came into the office to see if I was packing to go back east. I stayed, although I warned Mary and Stef, who were flying in for a visit in a few days, to expect aftershocks when they arrived. Even with the warning, they were unnerved when two aftershocks eventually occurred.

CTV competed with seven other teams to identify a viable highway or transit project somewhere in California and convince Caltrans to award it one of the four franchises. Over several months, we evaluated almost 100 projects before choosing State Route 125 in San Diego County as our candidate project. SR 125 was a proposed highway Caltrans had been planning to serve an area of pending development. The highway also would provide a link to the Otay-Mesa border crossing with Mexico where significant traffic growth was anticipated. Because Caltrans wouldn't have the finances to build SR 125 for over a decade, the highway would open to traffic some ten years sooner as a P3 than if its completion was dependent solely on Caltrans' funding sources.

We started preparing the SR 125 proposal, when several Ventura County landowners approached us about submitting a proposal to win one of the franchises for a toll road to facilitate development on their properties. They knew we planned to submit a proposal for SR 125, but were willing to pay CTV to submit a second application on their behalf. However, we decided we should focus on SR 125 and turned down the assignment.

While looking at the Ventura project, I lunched with Arthur Goulet, the Ventura County Public Works Director. He was my high school friend who originally had gotten me interested in civil engineering. I had lost touch with him, and it was a bit surreal as we spoke of what had happened to each of us in the 30 years since we last had spoken. Because his career mainly was in the public sector, I found it interesting to hear how he did his job over the years while typically reporting to non-engineers who were politicians in elected positions.

Jim Rozek had returned from Turkey and was helping CTV prepare the SR 125 proposal. One day we were working in a conference room, when he received a phone call from Ankara. The news was devastating — Andrew Dawson, the Parsons Brinckerhoff geotechnical lead on the Ankara-Gerede Motorway project, had died in a construction accident. He had been at the bottom of a high embankment viewing a slope slippage that had occurred the day before, when a bulldozer on the top caused some boulders to roll down the slope. One of the boulders hit and killed Dawson. Although the dozer operator was never charged with a crime, Rozek's replacement as Project Manager (call him Gene) feared potential criminal charges because of Dawson's death. Given that our work was substantially complete, I heard Gene opted to leave Turkey before the police completed their investigation.

I never found out if Dawson wore a safety vest, hard hat, and work boots for the site visit, although I knew he was very safety conscious. However, ever since the accident, I thought of him whenever I sent office workers to the field. Office staff aren't always as aware of construction risks the way field staff are, and I wanted to make sure everyone going to the field wore the proper safety gear to protect themselves.[19-1]

A monument was erected alongside the completed highway near the accident site as a memorial to Dawson. I visited the monument 25 years later to pay my respects. When I saw how weathered the monument had become, I successfully started the process to have a new memorial constructed.

CTV's financial projections indicated the toll revenues collected over the franchise's 35-year life would more than cover the costs to develop and implement SR 125. Such projections are based on many assumptions, including the rate of population growth in the corridor, the health of the economy, the cost of money, the ease of getting environmental approval, and material costs. In this instance, forecasted traffic demand at the border crossing also influenced project viability. Obviously, the reliability of projections becomes more and more speculative as one goes out 35 years.[19-2] For example, take population growth — who really knows with accuracy how much population of a region will grow or shrink over the next 35 years? Certainly, not elected officials who, to make themselves look good, always seem to claim positive growth in their jurisdiction, even when most demographers believe population likely will decline. Cognizant of all such uncertainties, we submitted our proposal for SR 125.

Bechtel, one of the teams competing with us for the four franchises, also chose SR 125 for its proposed project, while the other six teams chose different projects. Bechtel convinced the largest land developer in the SR 125 corridor to join their team, leaving us with a smaller developer on the CTV team. I came up with an idea that I hoped would cancel out Bechtel's move. During our oral presentation to the selection panel, I asked our developer, "Would you work with Bechtel if they won the franchise, and do you think the other developer would work with us if we won?"

He said, "Of course," to both parts of the question, which I anticipated would negate any advantage Bechtel hoped to gain. As our proposal and presentation were well received, CTV was one of the four teams to win a franchise, besting Bechtel and the three other teams.

Winning meant we would negotiate an agreement with Caltrans to establish the terms and conditions of the franchise. A new governor (Pete

Wilson) was to be inaugurated in two months, but one could only speculate if Wilson would be as supportive of the P3 approach as outgoing Governor George Deukmejian. For that reason, both parties knew we should sign the agreement before Wilson was sworn in. After several negotiations, there still were some major open issues with only few days left before inauguration. At that point, the parties agreed to settle all remaining issues at the next negotiation session.

Negotiations began at 2:00 pm in Sacramento and, after a short break for dinner, ended the following morning at 8:00 am. I sent out for pizza at 1:00 am so we wouldn't have to take a second break. CTV's negotiating team of four people consisted of two lawyers, my deputy (an engineer from Transroute) and me. By tag-teaming twosomes of a lawyer and an engineer, we always had two fresh people representing us while we resolved the final issues during this marathon negotiation. After Caltrans completed negotiations with the four winning teams, all agreements became public a few weeks later. I compared our franchise agreement with the other three and was pleased to find we did comparatively well.

CTV had spent $2 million to that point to win the franchise. The economy began slowing in early 1991, and Prudential Bache showed signs it would have trouble raising the funding required to turn a franchise into an operating toll road. A slowing economy also meant the real estate market would tank, and it would be unlikely the land developers we were counting on to donate property and money would be able to do so for years at best. Fewer new homes and jobs in the region meant there would be fewer cars and trucks on the roads than we assumed, resulting in less toll revenues. Moreover, my gut feeling was that our public pronouncements that we would obtain environmental approval in two years were optimistic. With all those uncertainties, I was not sanguine about the staying power of the four CTV partners for the long haul.

To alleviate my concerns, I proposed going to Bechtel, who had exhibited a strong interest in SR 125, and offering to sell them half the franchise for $2 million, the amount of CTV's investment to that point in time. With Bechtel's financial strength behind the franchise, I felt funding the project shouldn't be an issue going forward. Also, by recovering the money spent to date, the four initial CTV owners would be much more positive about the project's ultimate success. I never found out if my approach would have worked because Fluor was so opposed to working with Bechtel, their long-term competitor, that we never presented Bechtel with the offer. CTV then began to develop SR 125 on its own, and I relocated to San Diego to lead the effort.

I had planned to become licensed to practice engineering in California. Because California required professional engineers to pass a seismic engineering test, I felt I would require a refresher course as my structural engineering skills had waned and likely weren't strong enough. However, I was constantly on the move during my time in California and never had the chance to take a course or the test. Thus, others had to sign any engineering documents we prepared for SR 125.

After receiving the franchise, CTV's initial efforts primarily were building local support for SR 125 and preparing the environmental assessment. As I feared, the economy continued slowing and environmental obstacles began surfacing. It looked like it would be a long slog before the toll road could open for operation, and the euphoria of winning the franchise faded rapidly. Several months after we were awarded the franchise, it became obvious I had lost both Transroute's and Fluor's support as each started efforts to replace me as CTV's President. It was not surprising they wanted to fill the top slot with someone from their organization, as both originally coveted that position. Understanding staying was not an option, I resigned as President.

The CTV board selected a new hire of Parsons Brinckerhoff (call him Cliff) as my replacement having decided he was better than any other candidate. Upon meeting Cliff, I was surprised to determine he wasn't as qualified as I was. It made me wonder why Parsons Brinckerhoff's PIC on CTV hadn't fought harder for me to stay in place. The PIC claimed he was outvoted by the other three CTV partners, but I had doubts. As an aside, I admit to schadenfreude when Cliff lasted only a year at CTV.

This was perhaps the low point in my career. Now my family would have to rethink plans to move to California, while I had to find my next assignment. Even though Lammie and Rubin, respectively Parsons Brinckerhoff's President and Chairman, indicated their support for me, I speculated whether I would have to leave the firm.

I moved back to Orange, California, while the word went out firm-wide that I was available for another project. Before long, I helped win my next project, which ended any concerns I had about my value to the firm. Meanwhile, with SR 125's future increasingly in doubt as the economy worsened, my going to work on a new project was looking very favorable. In fact, it was ten years before Parsons Brinckerhoff was able to make a profit from its investment by selling its ownership interest to an investor/developer (call it ID Inc.). It took another two years before ID Inc. achieved final environmental approval to build the toll road — totaling twelve years after CTV obtained the franchise, instead of the originally contemplated two years. My instinct that things may not go as hoped for had been right this time. Further, it would have been much less interesting for me to have treaded water with the SR 125 project for those twelve years.

ID Inc. opened the toll road, named the South Bay Expressway ("SBX"), to operation in November 2007. However, ID Inc. went into Chapter 11 bankruptcy in March 2010 when revenues could not cover expenditures and liabilities, including large contractor claims and related legal costs to contest the claims. SBX's history underscores that earnings and revenue forecasts are based on assumptions and cost estimates that are highly sensitive to variation.

Twenty years after I resigned as CTV's President, an opportunity to work once again on SR 125 came as I retired in 2011 from Parsons Brinckerhoff. The creditors and banks that financed the road's construction (collectively the "Lenders") were preparing to assume ownership of SBX as it emerged from ID Inc.'s bankruptcy. The reorganized company, South Bay Expressway LLC, would run the toll road, and the Lenders were seeking candidates to serve on the company's board.

Chuck Fuhs, a colleague from Texas, told me about the opening and suggested I apply. I jumped at the opportunity to regain pride on a project where I had been compelled to resign. Given my first-hand knowledge of the project and board experience with Parsons Brinckerhoff, ASCE, and others, I felt I offered skills and experience no one else could match and submitted a résumé.

My references were Wayne Klotz, a past president of ASCE, and three former non-employee Parsons Brinckerhoff board members: Rodney Slater, a past US Secretary of Transportation; Dick Ravitch, a past Lieutenant Governor of New York; and Tom Opladen, an active member in the National Association of Corporate Directors. In April 2011, the Lenders appointed me a board member representing the U.S. DOT, who had provided a TIFIA loan to the original owners. The Transportation Infrastructure Finance and Innovation Act, or TIFIA, program provides credit assistance for qualified projects of regional and national significance. The position required two or three days a month of my time.

The Lenders had no desire to own and operate a toll road and immediately started looking for someone to purchase the franchise from them. The San Diego Association of Governments, commonly called SANDAG, offered to purchase the toll road, and after negotiations and due diligence, the sale was consummated in December 2011.[23] Over the next few months, the Lenders moved to liquidate the now superfluous LLC, and my involvement as a board member ended in April 2012.

The previous year of serving part time on the SBX board showed me that I no longer needed to be working to be content. At that point, my retirement from paid work began, although I continued volunteer activities for ASCE and CCNY.

---

[23] The sale was a very good outcome for U.S. DOT, as the TIFIA loan was fully repaid in 2017.

CHAPTER 20

MISCELLANEOUS RESPONSIBILITIES

Even when busy, I found it difficult to refuse requests for help and frequently took on tasks unrelated to projects I was managing. As I became more experienced, I realized it's better not to accept an additional assignment when it would impact my ability to manage my current project.[20-1] (Lammie often used the analogy of adding another spinning plate on a stick to those you already are spinning.) I learned to ask my supervisor to decide whether my current effort or the added one has the higher priority, whenever I felt I couldn't perform both successfully.

On the other hand, there were frequent periods when my project was slow-moving or I was between assignments, and it was natural to worry if upper management will think I've become superfluous. That's when I'd look for something to do and didn't just wait for someone else to find work for me. Such slow periods were a good time to seek a peer review assignment, a short-term assignment on someone else's project, or a non-project task to undertake both for personal benefit and the good of the organization. This chapter provides examples of assignments I was able to accomplish while in a slow period.

Acquisitions:

Periodically, Parsons Brinckerhoff considered whether to acquire another firm, and I was involved with the due diligence efforts evaluating several potential acquisitions. One effort was related to the potential purchase of an 80-person engineering firm in Spain (call it Ingeniero), where I was responsible for due diligence of the firm's engineering capabilities. I went to Seville and interviewed Ingeniero's engineering staff to assess their capabilities and weaknesses. My Spanish was at best adequate, so sometimes I used an interpreter. While there, I asked if I could speak with someone from their largest client, the national highway agency, to learn what the agency thought about Ingeniero's engineering skills. I was told that speaking with the agency would be a problem, because Ingeniero hadn't told the agency it may be sold. Further, Ingeniero's senior managers went on to say there was no reason to visit the agency, as I could believe them when they said they had no problems with their main client. By saying there was no need to verify their statements, Ingeniero was asking me to trust them when I barely knew them. My natural skepticism kicked in, and I only probed deeper.

Parsons Brinckerhoff had retained a Seville firm to advise us and represent our interests during the due diligence process. Oddly, the advisory firm spent most of its time trying to convince me why we should hurry to buy Ingeniero. I began to suspect the advisory firm's loyalty, as it only said overly positive things about Ingeniero. The whole state of affairs was a red flag to me. When I finished what I could do in Spain and returned stateside, I told Parsons Brinckerhoff's due diligence team leader that Ingeniero's engineering skills seemed average at best. I also said we (a) should try to find out why no one in their main client's office could be used as a reference and (b) should be wary about advice from the advisory firm.[20-2] Informed that others would address my concerns, I moved on to resume efforts unrelated to the acquisition.

A few months later, I heard we bought Ingeniero. I asked what their main client had said about the firm and was told we hadn't spoken to them. Ingeniero pushed us to make a decision to buy them, and we yielded-a big mistake. Over the coming months, we learned Ingeniero's main client was disappointed in the firm's work and would not be giving it other assignments in the future. This drop-off in work resulted in numerous staff lay-offs with associated high severance costs to Parsons Brinckerhoff, something we would have anticipated with proper due diligence defining the facts. After acquisition, Murat Tasar (with whom I'd worked in Turkey) was tasked with coordinating Ingeniero into our business systems as our Integration Manager. Tasar told me he learned the advisory firm had worked behind the scenes for Ingeniero and not solely on our behalf.

The acquisition of an engineering firm (Keystone/Avid) in Albuquerque, New Mexico, was more successful. In that instance, I managed a due diligence team of seven, with different team members focused on the financials, engineering, project management, legal, operations, marketing or human relations. I was pleased with the team's efforts and felt the final report was thorough, covering the good and bad points of the firm and the potential benefits and risks in a detailed manner. I've maintained a friendship with Andres Viamonte, one of the three major owners of Keystone/Avid, even when he left Parsons Brinckerhoff three years after the acquisition (which I had anticipated in my final report). And the New Mexico operations were still doing well when I retired some two decades after the acquisition.

Project Management Handbook:

While I was in Hong Kong on CKR, Parsons Brinckerhoff created a Project Management Manual. It was a very thick, three-ring document, covering project management issues in copious detail and describing the numerous, internal firm procedures and forms. I literally lugged the manual home one weekend, and it took me sixteen hours on Saturday and Sunday to read it. I thought it illogical to expect most Project Managers would read the whole document, and if they did, that they would remember everything in it. I had learned from my ACEC peer reviews that it's not realistic to expect that everyone both understands and is complying with every policy in force.

Before long, managers of small projects complained the manual was too exhaustive and requested something simpler. Lammie asked me to create a handbook version of the manual, which I was happy to do. After all, I, too, wanted something simpler.

Metrication:

In the 90s, a new regulation mandated all federal agencies go metric in ten years. I supported that move because I felt one makes fewer blunders in the metric system, where the difference between millimeters and meters is 1000, then in the Imperial, or English, system, where the difference between feet and yards is a factor of three. In other words, you're much more likely to confuse twelve square feet with twelve square yards than confuse twelve square millimeters with twelve square meters. Nonetheless, when you abandon one system to work in another, you lose many of your sanity checks and have to create new cues to use to spot problems.

Realizing metrication would be confusing to most Americans, I proposed Parsons Brinckerhoff form an internal committee to help its domestic staff and clients go metric. Lammie agreed and appointed me to lead the committee. I identified an expert in every discipline who had worked in both the metric and Imperial systems and tasked them to develop training aids for their discipline. Next, I created wallet-size cards with basic conversion factors and pocket metric scales as handouts to clients. I envisaged many clients hiring us to convert their standards and specifications to metrics.

We were almost ready to roll out the program, when Congress folded under political pressure and rescinded the requirement to go metric. It seemed the various states, municipalities and counties around the country were not willing to pay to change all their speed limit and destination signs and pressured their elected federal officials to oppose metrication. Nevertheless, I felt I had been appropriately visionary in looking for a way to be ahead of the curve.[20-3]

Review of Firm-Wide Policies:

In 2008, Keith Hawksworth became Parsons Brinckerhoff's CEO and asked me to go around the company and assess which firm-wide policies were functioning well, which should be modified, and what new policies were required. I visited ten offices in the U.S. and eleven international offices to get feedback from a cross-section of personnel in the firm. I randomly selected and interviewed over 300 employees of different job functions, grade levels, and disciplines in the 21 offices. I codified the most important comments I received and suggested approaches to deal with issues I noted. There was one comment from the many interviews that still resonates with me:

> As the only revenue an engineering firm makes comes from the actions of its Project Managers, upper management should first ask how it helps Project Managers before reorganizing or instituting a new policy.

Chapter 21

TWO NOT SO SUCCESSFUL PROGRAM MANAGEMENT ASSIGNMENTS

I'd previously mentioned I lost money on a toll road project in New Jersey. I subsequently managed other projects and programs that were unsuccessful for one reason or another, and the two programs detailed in this chapter stand out in my mind. It may just be a coincidence, but both these programs also were in New Jersey.

I grade my efforts on the two programs as B+ for financial, since they were profitable, but D for client relations. While I did what I could to make both successful, they were highly stressful assignments with little the firm could use as a reference when pursuing other projects. I learned from the two programs how software-hardware development is highly risky and about the need for an early warning system to identify if a project is headed for trouble. Also, the programs reinforced how critical it is to have a good document control system. Not a full payback for all the headaches, but as mentioned earlier, we learn more from failure than success.

Regional Electronic Toll Collection:

The Regional Electronic Toll Collection ("ETC") Program in 1998 involved an electronic toll collection system and a fiber optic system on 435 miles of toll roads in New Jersey, New York and Delaware. The client was an association of five public toll agencies: New Jersey Turnpike Authority (the lead agency), Port Authority of New York and New Jersey, New Jersey Highway Authority, South Jersey Transportation Authority, and Delaware DOT. The owners had awarded a contract for $500 million to a firm to design, build, operate and maintain ("DBOM") the program. Parsons Brinckerhoff's program management efforts on behalf of the owners included contract administration, review of the design, and construction inspection and quality assurance of the DBOM firm. I was the Program Manager, and Mel Stein was my Manager of Project Controls, the same roles we had on Westway. As on the AC/BC project, I used an engineer from the PGN to help during start-up.

The five toll authorities' alliance was a major coordination problem, as all decisions had to be unanimous, and each agency was used to doing things its own way. Thus, even simple decisions took an inordinately long time. Another problem was that the DBOM firm was grossly under-financed. In fact, several months into the program, the firm went under, only to be replaced by another firm that also had financial problems. The two DBOM firms' inadequate financing together with the lack of timely direction from the owners, meant every hardware-software milestone was missed, and hardware and software development failed to advance as expeditiously as it should have. As a Highway Engineer, I never had been involved with high-tech hardware and software development, but felt this project had more than enough for a career. (Little did I know my next project also would have major hardware-software problems.)

The ETC program's many software development issues reminded me of when Wally, the underachiever in the "Dilbert" comic strip, stated, "Progress is difficult to measure in the software realm. You could measure the lines of code I produce, but that would reward inefficiency."

Wally leaves us to wonder if one can ever tell if software tasks are ahead or behind schedule. Experience led me to conclude that hardware-software development is more an art, than a science. Such development rarely finishes on time or budget, either because end-users can't define what they want well enough to the hardware-software developers to get what they need, or developers don't prepare realistic estimates of how much time and money are required to complete the work. Regardless of whether the problem is the end-user or developer, my recommendation on hardware and software development efforts is to have substantial schedule and cost contingencies, perhaps even two or three times initial estimates.[21-1]

Several months into the project, a serious incident occurred one night when the DBOM firm's contractor laying fiber optic cable cut a major pipe providing a third of the water to Jersey City, a city of over 250,000 people. Our inspector at the site immediately notified his supervisor, who came to the site that night, but failed to contact our client or me. First thing in the morning, the media contacted our client's Project Manager (call him Carl) who then called me to ask what happened. I had no idea what he was talking about, but said I'd get back to him, thinking it looked like I wasn't managing anything. Eventually, we learned the primary cause of the incident was the water utility company failed to mark the waterline's location in the field, which meant the contractor was unaware there was a line to avoid. From all the negative comments we received from the client in the following weeks, it bewildered me as to why they apparently preferred blaming my firm and me for the waterline cutting, rather than the utility company who didn't mark the waterline's location.

While the primary fault was others, we had our own communications failure. Namely, I don't want my client to call me, saying their supervisor asked about a problem they know nothing about, as it's always better for them to hear bad news first from me than from others.[21-2] Never wanting to be in that position again, I followed Jenny's approach and created an early warning system, which I named the "30-Minute Rule:"

> Everyone has 30 minutes to tell their supervisor and client counterpart when something goes wrong or significant happens. You don't have to know how to solve the problem at that point, but you honestly can say you've started the process to find a solution. This rule applies up-the-chain until both Parsons Brinckerhoff's Project Manager and the client's Project Manager are notified. If you can't reach me after a good faith effort, take whatever action you believe warranted, and I'll support what you do.

As the program's slow progress continued, Carl became so frustrated by everyone's performance that he lost control of meetings. Too many meetings evolved into shouting matches and finger-pointing exercises with four-letter curse words by Carl thrown in for emphasis. After one especially dysfunctional meeting, I asked to meet privately with him and suggested he change his approach as senior engineers and managers would respond better to more measured criticism. Carl thanked me for my candor, but said he couldn't change how he ran the meetings as he felt his approach was what was required. Before long, the project became very draining on team members (whether from the DBOM firm, one of the five toll authorities or Parsons Brinckerhoff), and it seemed as though everyone was trying their best to get off the project.

As the ETC project slowly dragged everyone down, I started looking for my own exit strategy so I could shift to the Motor Vehicle Inspection program (discussed in the next section), which was getting ready to start after a year's delay. As the client lead agency wanted to replace me anyway, mostly related to the waterline cutting, Greg Soriano took over for me as the ETC Project Manager.

New Jersey Motor Vehicle Inspection Program:

I started as Project Manager on the Enhanced Motor Vehicle Inspection/Maintenance System program for the New Jersey Division of Motor Vehicles ("DMV") in late 1998. George Oliger had been the pursuit manager for the project, and the DMV ranked our proposal first. The interview had gone well, and we won the assignment somewhat handily.

This DBOM program involved implementing an enhanced automobile inspection and maintenance system in 35 existing and proposed state inspection facilities. Our efforts as program managers included contract administration and quality assurance of the DBOM firm who had a contract in excess of $500 million (in 1998 dollars) with the DMV. Quality assurance tasks involved (a) reviewing final design of the facilities, (b) reviewing hardware and software development of the in-lane inspection equipment and the data management system, and (c) monitoring construction at the facilities. Once again, an engineer from the PGN helped during start-up.

Federal Clean Air Act Amendments of 1990 mandated that New Jersey have the vehicle inspection system working by December 13, 1999 or risk losing all statewide federal highway funding. My Deputy Project Manager, Beth DeAngelo, bought a countdown clock, which we set to the scheduled opening date to remind us that an hour lost can never be recovered. DeAngelo was an outstanding colleague, with the ability to smooth over the many rough edges that arose on this program. She showed me that I could learn from someone three decades younger than I was, and she helped make me a better manager.

The DBOM firm had several subcontractors including a construction contractor, a materials testing firm and a construction inspection firm. We periodically observed the DBOM firm and its subs and frequently were disappointed in the level of effort and work quality we saw at the 35 facility sites we randomly visited. An extreme example was when our senior observer, Fred Colacino, went to a construction site and saw the material inspector's signed slip approving the concrete slump test results that day. However, while Colacino was reading the approval slip, the first concrete truck was only just arriving at the site. It was impossible for a test to have been performed as no concrete would have been at the site when the test slip was signed — **the slip approving the concrete was a complete fabrication**! Colacino conducted his own test on the just-arrived concrete and found it slumped double what it should to be approved. That meant inadequate concrete would have been poured in the foundation had Colacino not showed up that morning at that site. And the fact he did show up was but a chance occurrence, as it just happened to be the random site he visited that day. We reported our findings to the client's Project Manager, call her Susan. Susan surprisingly only slapped the wrist of the testing firm and didn't hold the DBOM firm truly accountable for faulty oversight of its subcontractor's performance.

The DBOM firm also was responsible for developing the program's hardware and software, and our role included reviewing what was developed and offering opinions on adequacy. To help us in this effort, we retained a California-based subcontractor, Sierra Research Inc. ("SRI"), widely recognized in the industry as a leading specialist in the design and evaluation of advanced emissions inspection technology. From the start, SRI doubted the adequacy of the equipment to be furnished, the sufficiency of the training program for the workers in the stations, and the likelihood schedules could be met. For example, while temperatures often fall below freezing in New Jersey winters, the DBOM firm failed to specify equipment that would operate in that environment, even after we pointed that deficiency out to them. Repeatedly, the DBOM firm failed to make an acceptable submittal of a key interim deliverable. Nonetheless, Susan would permit them to move to the next phase.

For some reason, Susan acted overly sympathetic to the DBOM firm who continually complained that we were too strict in how we interpreted their efforts. At first it wasn't apparent, but Susan gradually froze us out of tasks where we had opportunities to critique the DBOM firm's work. For example, she stopped notifying us about review meetings during which she approved work efforts, even when she probably knew we likely would reject some efforts. Also, Susan eliminated our role of reviewing the DBOM firm's invoices, and she approved payments for incomplete efforts. A cynic might say that was because we said we would recommend against payment for any work not in accordance with the contract.

As tension with Susan ratcheted up, I became apprehensive she might go to her supervisor and ask that Parsons Brinckerhoff be removed from the project. I knew we didn't deserve to be terminated and decided to try to pre-empt that possibility. I went to Susan and offered to resign from the project, saying it seemed she was disappointed in my performance. My goal was to let her blame me, but not Parsons Brinckerhoff. After all, if she blamed the whole firm, the client may terminate the contract with the firm and our project staff would be out of work. My offer appeared to catch her by surprise, and she said I should stay and continue in my role. After that encounter, I no longer sensed that we might be terminated, although she still often ignored our advice.

With a few months to go, it was apparent to SRI and us that it was highly unlikely the scheduled opening date could be attained with a working system operated by a suitably trained inspection station workforce. Lammie had trained me to act decisively on negative findings;[21-3] and to that end, I proposed a series of what-if meetings.[21-4] Attendees at what-if meetings would strategize **WHAT** to do to minimize schedule slippage and, **IF** we determined the mandated completion date would be missed, how best to convey the situation to the federal Environmental Protection Agency such that they would grant an extension to the completion date. Good risk mitigation strategies came out of these meetings, but it was all for naught as the meetings were anathema to Susan. After attending the first one, she characterized what-if meetings as negative thinking and refused to attend any more.

We had the strategies, but no means to implement them. The situation was very confusing, as I couldn't recall previous incidents where a client wouldn't accept our ability to question the adequacy of a pending major deliverable or suggest cures for potential schedule slippages. We continued speculating as to what motivated Susan, a Professional Engineer, to reject our comments and suggestions, while always accepting the DBOM firm's version of events.

Given the tenor of events, I felt compelled to write Susan a detailed letter summarizing our concerns and getting them on the record in no uncertain terms. To me, documenting and explaining our opinion was the proper and ethical thing to do; merely sitting back and collecting our fee would not suffice to meet our professional obligations.[21-5] It may have been prescience on my part, but I strongly believed we would be blamed for everything if things didn't go well. I felt the need to manage that risk by using the letter to get the client to accept there was a very high possibility things would go poorly and actions should be taken in that anticipation. I also believed I had to go over Susan's head, because I sensed she would ignore the letter. Thus, I gave her supervisor, the Director of the DMV, a copy of the letter saying we believed the program's success was in jeopardy if things continued the way they were. The Director said he appreciated our concerns and would bring them to his supervisor, the New Jersey Department of Transportation ("NJDOT") Commissioner. Two weeks later, he reported the Commissioner had said nothing would be gained by making changes at this late date and that he was relying on Susan to finish the job. I was very disappointed in this decision, but we soldiered on as best we could. In hindsight, I should have been even more decisive and continued going over heads, until I had elevated our concerns to the Governor, Christine Whitman.

In the weeks before the opening, we sensed worker training was marginal at best and based more on hope than smooth test runs, but the DBOM firm and Susan said all would work out well. The DBOM firm did make the mandated opening date, but within a month, significant quality issues surfaced when the temperature fell below freezing, and the whole system crashed as we said it might. The DMV program had very high, statewide visibility, and the media had a field day with the system shutdown.

We met the NJDOT Commissioner to describe what we felt went wrong and offered suggestions for recovery. The governor had an upcoming press conference to discuss the program, and we expected the Commissioner would pass our recommendations on to her. Two hours later, the governor was on television saying Parsons Brinckerhoff (not the DBOM firm) had failed. We were stunned our client publicly blamed us for not providing appropriate oversight of the program and threw us to the wolves even though our predictions of quality failure had proven true. The trust we had in some senior government officials to treat us objectively was misplaced as they looked for a scapegoat to protect their reputations; they made it appear our warnings of potential failure were sanitized and not clearly conveyed. Some in the media went for sensationalism and blamed us without hearing our version of what happened, but the major newspapers knew Parsons Brinckerhoff well enough to tell fair stories while waiting to learn what really occurred.

Because we had documented our concerns in writing to the client, our records enabled us to respond quickly to the governor and the media. The paper trail confirmed the timeliness and appropriateness of our actions and documented Susan's lack of proper concern during project implementation. Eventually, I testified under oath three times before state investigatory panels, where I had an opportunity to explain our actions. When strongly pressed for my opinion why Susan did what she did, I said I didn't know, but assumed it was ineffectiveness and not corruption. Because I knew testifying would be intense, I had Parsons Brinckerhoff provide me with outside counsel to help me avoid inadvertent misstatements and assist me in general.[21-6]

The investigatory panels' final reports justified our efforts. Once the reports came out, the state's largest newspaper, *The Star Ledger,* conveyed the facts in an article under the headline: "Car-Test Warnings Not Sanitized: Explicit Memos [*from Parsons Brinckerhoff*] To State Spoke Of Equipment Trouble, Long Waits."

While neither the governor nor the commissioner ever apologized to us, Susan was relieved of her position. As happened on the ETC project, we were splattered by mud from the mistakes of others on the DMV project. However, by acting decisively on negative findings and having good document control practices on DMV, we minimized the damage to us.

CHAPTER 22

MANAGING A MEGA-PROJECT IN HOUSTON

I pursued two projects as Parsons Brinckerhoff efforts on the DMV project neared completion in 1999. One was a rail project, where I would be based in London, England, and the other was a highway project in Houston, Texas, and I would manage whichever project Parsons Brinckerhoff won first (if either). While London was my preferred assignment, running a mega-highway project in Houston, also sounded like a good posting.

The Houston client, the Texas Department of Transportation ("TxDOT"), was the first to select a firm. TxDOT wanted to retain a general engineering consultant ("GEC") on the $2.7 billion (in 2000 dollars) reconstruction of 23 miles of the Katy Freeway, IH-10. For most of that length, the reconstruction consisted of adding seven lanes to an existing eleven-lane facility. GEC services involved technical and administrative management of the design, including reviewing and coordinating ten section design consultants ("SDCs"). The SDCs would be under contract to TxDOT and would prepare designs for nine major construction contracts. Each contract can be thought of as a project, while all the contracts combined are a program. Treating the Katy's nine related projects as a program was unprecedented for TxDOT, as they had never used a GEC or Program Manager before.

Parsons Brinckerhoff was not known in Houston for preparing highway design packages, and we realized we needed a game-changer to win the assignment. We presumed our competitors would offer a local Project Manager who was known to TxDOT and strong in design, but lacking GEC experience. Our plan was to convince TxDOT that they would be taking a chance using a local Project Manager from one of our competitors for the effort. Prior to TxDOT advertising for a GEC for the project, we invited TxDOT's Project Manager to see how program management was performed on another mega-project: the Woodrow Wilson Bridge connecting Virginia and Maryland. That visit showed

TxDOT the risks faced if the Program Manager for the Katy program lacked the required coordination and managerial skills. By proposing me as the Project Manager, the game changer was that of all the firms competing for the GEC role, only Parsons Brinckerhoff was offering an experienced, world-class manager of GEC/program management assignments.

The interview went very well, and Parsons Brinckerhoff won the Katy assignment in late 1999; I turned the DMV project over to DeAngelo to complete the firm's few remaining obligations. By the way, ultimately, the firm also won the London assignment, and I was told it became an extremely testing one. In hindsight, I was very fortunate the Houston pursuit was decided first.

In nine of the previous ten years, the public ranked "transportation congestion" as Houston's worst problem. Everyone expected widening the Katy would relieve that problem, and the reconstruction had about 99% approval from the local citizens. I can't imagine another major American city supporting so much new highway pavement. With the public behind the project, elected officials and politicians followed: Texas Governor Rick Perry, the U.S. Secretary of Transportation Norm Minetta, the Chair of the U.S. House of Representatives' Transportation Committee Don Young, and Houston's Mayor Lee Brown all came to public meetings to support the Katy widening. The local Congressman, Rep. John Culberson, visited our office three times to make sure the project was on schedule. He was impressed that we had a countdown clock in the conference room; I'd brought it from New Jersey and had set it to the first milestone date to make sure everyone knew our next target.

The vast majority of those working on the Katy project (whether from TxDOT, Parsons Brinckerhoff or an SDC) had never been involved with a GEC-led mega-project before, so my initial efforts included counseling those on the project what to expect and convincing them they should buy into my program management approach. I felt both TxDOT and the ten SDCs would be apprehensive that we would try to make SDCs look bad to make ourselves look good, as every SDC was our traditional competitor. Thus, I expected some SDCs would try to defuse our efforts by going directly to TxDOT to complain about this or that and saying we were unreasonably pressuring them (similar to what the DBOM firm on the DMV project did). Nevertheless, I had to get the SDCs working productively with minimal bickering.

Because I wanted TxDOT to allow us to organize the program in ways we'd found successful in the past, I reiterated to them that we knew more about the role of a GEC than they did. I reminded TxDOT they retained us to take the burden of managing ten SDCs off their backs and should avoid hampering that objective. I said my approach is to help everyone succeed, and that way the whole program succeeds. Fortunately, I persuaded TxDOT to accept that SDCs would write and speak to TxDOT only through us, and TxDOT would rebuff SDC attempts to circumvent this approach. Once TxDOT accepted those recommendations, everything worked as planned. Before long, even the SDCs realized we were there to help them. It was the combination of our knowing what had to be done and TxDOT's willingness to put its faith in us that resulted in the Katy project becoming successful.

Because the role of GEC was new to our own local staff, I relocated several Parsons Brinckerhoff employees to Houston with experience on large projects who, therefore, could do what was needed with minimal guidance. It was an organization where more than half the GEC staff were new to Houston, which meant many personnel hadn't worked with one another before. I could feel tension building as everyone sought to define their role in this new organization. Before bad practices became set in place, I asked the firm's Human Resources Department to hold a teambuilding workshop to transform the organization from a group of individuals to a cohesive team working well together. The training calmed the waters enough that most everyone began pulling in the same direction.

Things, of course, weren't perfect; for example, one woman felt unduly picked-upon by some of the men. I looked into the situation and felt it was not gender bias, but aggressive teasing of a co-worker. However, because it involved female-male interactions, I had Human Resources hold a training session on gender bias/sexual harassment for the staff. A year later, the woman quit the firm, which said to me the teambuilding issue may not have been dealt with as well as I wanted. I now realized both teambuilding and gender bias/sexual harassment training can never stop.

When I relocated to Houston to be the Katy Project Manager, I also agreed to serve as acting Houston Area Manager until a permanent Area Manager was appointed. Winning the Katy project also meant we had to move to much larger quarters to accommodate all the new staff. From the Harvard-MIT seminar I'd taken, I'd learned office size and location and the presence of a window or door matter to many people.[22-1] I had myriad other important things to deal with, yet knew morale could fall if I failed to meet everyone's expectation. I calmed staff by letting everyone choose their own office in turn, starting with the most senior employee. That approach worked as not a single person complained as each got the office they deemed best out of those still available. I did notice one new hire never personalized his office; he didn't place any photos, licenses or diplomas on his desk or walls. I presumed he wasn't planning to stay long, which proved prophetic when he quit before too long.

Cell phones were ubiquitous by 2000. At every meeting, someone's phone would go off and interrupt the flow of the discussion. Because I tried to limit meetings to one-hour, delays from attendees answering phone calls were disruptive. Knowing callers could reach attendees in an emergency via the firm's main number, I created a meeting rule: Put your phone on silent or vibrate; otherwise, you give $5 to charity when your phone rings during a meeting. After a few charity donations, the problem was solved.

SDCs had to meet milestone delivery dates for 30%, 60%, 90% and 100% submittals. I was concerned we wouldn't have sufficient hours in the agreement for reviewing SDC submittals and would exceed our budget. We started negotiating the number of required hours with TxDOT by estimating the number of sheets per submittal and the number of hours to review a sheet based on four milestone submittals per SDC. I told TxDOT it was logical that portions of the four submittals would be rejected, and we probably would review an average of five submittals, not four, from each SDC. TxDOT agreed, and the budget for the fifth review gave us the sufficient contingency we required.

Making sure SDCs made those four submittal dates was important both for Parsons Brinckerhoff to stay on budget and for the project to meet its planned open-to-traffic by December 31, 2008. To mitigate against schedule slippage, I had TxDOT agree to add a clause to the SDC agreements which required SDCs who missed a delivery date to pay Parsons Brinckerhoff's overtime costs. Eventually, all dates were met, as you can imagine no SDC Project Manager wanted to go to their supervisor asking for a check to reimburse a competitor because a date was missed.

The Katy schedule was aggressive, but not unrealistic. Early on, one of my deputies came and said an SDC submittal date would be missed by a few days, and that TxDOT's Project Manager had said it was okay. I said it wasn't all right with me, and the SDC had better make the due date or pay our overtime costs. The SDC made the date. I felt it was important to never ease up the pressure, or, before long, the schedule becomes meaningless. My objective was to get to the point where, if asked what concerned me the most, project staff would say, "Staying on schedule."

It must have worked, as the widened Katy eventually opened to traffic more than two months early in October 2008. TxDOT got rave reviews from the public, politicians and media, and we were praised by TxDOT. After all, how many mega-projects do you know that finished early?

To control costs and schedule, I asked each staff meeting attendee to identify anything with a cost or schedule implication. Treating the Katy as a program with a strong emphasis on cost and schedule control ultimately were the keys to delivering improvements of Katy's magnitude as rapidly as possible with the least disruption to motorists during construction. The success of that approach helped the Katy become transformational for TxDOT, whereby they expanded our role from general engineering management into program management and contractual management. Once we had the opportunity to demonstrate we could remove those burdens from them, TxDOT confidently added more efforts into our scope. For example, partway through the project, a new Federal Highway Administration requirement mandated that TxDOT prepare an annual financial plan for the Katy project, and TxDOT assigned preparing the initial financial plan and its annual updates to our scope.

190 THE ENGINEERING IS EASY

Four of the 18 proposed lanes on the widened Katy Freeway would be managed lanes in the median (a managed lane is one that is restricted in some way, such as truck only, HOV only, tolled, or reversible). Today, at its intersection with Beltway 8, the Katy Freeway is 26 lanes wide and is reputed to be the widest highway in the world. The 26 lanes include twelve main lanes (six in each direction), eight feeder lanes and six managed lanes. Tongue in cheek, I'd remark at how beautiful that sea of concrete was, and that it was environmentally wonderful because all that concrete meant there were no disease-carrying mosquitos or ticks and little pollen to bother anyone with allergies.

Several months into our efforts, the Harris County toll agency ("HCTRA") offered to pay TxDOT for the right to finance, build and operate the four managed lanes as a toll road. TxDOT's Houston office wasn't savvy about toll roads and accepted my offer to prepare a white paper describing all the issues so they could negotiate a fair agreement with HCTRA. A key participant in the effort was Chuck Fuhs, who had coined the term "managed lane" for the profession and was the firm's expert on the topic. TxDOT was very pleased with the white paper we prepared, further enhancing our credibility with them.

While working on the Katy project, I peer reviewed a project in South Carolina. The project staff had created some user-friendly, internal reports for the project team. After reading these reports, I immediately wanted to create a similar report for the Katy project, but one intended for external distribution, rather than only internal. The report would be something useful we proudly could give to the media, elected officials and public citizens. Such a report would not be sugarcoated, but informative of what was happening.

I came back to Houston and had staff create a quarterly report, written without technical jargon and filled with photos and colorful charts. Each chapter had a traffic signal symbol to indicate what was proceeding on target (green), what was lagging (amber), and what was in trouble (red), so readers could quickly grasp project status. As I anticipated, TxDOT liked the first report so much, they added preparation of future reports to our scope. This instance substantiated my previous comment that performing peer reviews helps the reviewer as much as the reviewed.

Two years into the project, TxDOT had a statewide financial crisis and was running out of money for that fiscal year. It was obvious to me that eventually TxDOT Houston District would defer non-critical SDC efforts to cut back on current expenditures, and I reduced our GEC staff commensurate with the expected SDC reduced work efforts.[22-2] When TxDOT got around to directing us to cut our costs, I said we anticipated their request and had already reduced staff. That helped TxDOT realize we weren't a firm that unnecessarily ran up costs, and it legitimized future requests when we needed funding for new assignments.

The TxDOT Houston District had always performed its own construction inspection ("CI") without retaining a CI consultant. Nevertheless, I sensed an opportunity for Parsons Brinckerhoff to provide CI services on the Katy and came up with a way I hoped would make it happen. First, I had our CI staff prepare a staffing plan for inspecting each of Katy's nine construction contracts, and then layer the nine staffing plans on top of each other to develop a cumulative estimate of Katy's CI staffing needs. I shared the data with TxDOT saying I wasn't sure if they realized how many CI staff they would require in the peak construction years and these data provide the information to plan the staffing. Of course, I also said we were willing to help if they didn't have enough internal staff in those peak years. They thanked me for advising them of what they faced and enabling them to start their planning for CI staff. And when the time came that TxDOT required more field staff than they had in-house, we provided twelve inspectors and office engineers for two years.

TxDOT purchased 20 extra feet of property on the north side of the Katy for a project utility corridor. Their intent was to move all existing utilities out of the way of the widened Katy and relocate them into designated horizontal and vertical locations in an underground utility corridor. Seven utilities, including a high-power transmission line, would have to fit into the corridor's trench. It occurred to me that if the first few utilities were constructed even slightly off their designated locations, there would not be room for the last utilities to fit into the trench. I explained to TxDOT that they should have inspectors and surveyors check as each utility is constructed or risk finding out there isn't space for that last utility. Once again, I offered to provide inspectors on behalf of TxDOT, which eventually they accepted. It became win–win as all utilities fit into the trench, and the firm averaged five inspectors for two years in that effort.

We had many people going to the site, either walking the alignment during the design stage or performing CI services. Remembering Dawson, I knew we had an obligation to train our staff what to expect to help keep them safe, and I opted to provide them with the safety training required to become certified per the Occupational Safety and Health Act (commonly called "OSHA").[22-3] When the dates were set for the training, I called TxDOT and asked if they wanted to send some of their staff to be trained and become OSHA certified. TxDOT jumped at the chance and was grateful we provided the opportunity to their employees.

Project risk management had become a firm-wide standard as we started the Katy project. As I looked at what others were doing, it seemed many Project Managers did risk assessments very perfunctorily. They considered risk management a burden they were required to do by company policy and not something of value to them. It appeared to me they were comfortable dealing with tangible risks, but not abstract ones. I, on the other hand, felt risk management must be treated seriously.[22-4] For example, there not only are project financial risks, but also risks to client relations. On that basis, I started risk assessments by trying to identify what could happen such that my client may never want to hire our firm again. I knew that if we were to significantly underestimate a project's cost, ultimately the media and elected officials would blame the client who then would blame the firm-and, naturally, when the firm is blamed, I, as Project Manager, will be blamed by both the client and my supervisor.

A book dealing with the risks of planning and constructing mega-projects entitled *Megaprojects and Risk: An Anatomy of Ambition*, by Bent Flyvbjerg, et al, criticized the cost estimating of many mega-projects (defined in the book as multibillion-dollar infrastructure developments). The publisher stated that book's central theme is that promoters of multibillion-dollar projects may misinform lawmakers, the media, and the public in order to obtain construction approval for mega-projects. The book presents evidence that studies used to justify transport mega-projects typically underestimate costs and overestimate benefits, sometimes by orders of magnitude. While I feel to some extent the authors let pre-conceived biases influence their findings, the history of underestimating mega-project costs makes it a risk one can't ignore.[22-5]

Part of the problem is some cost estimators estimate what they are asked to estimate without considering what the project is likely to encounter. As an example, if I asked estimators for the cost of 500 feet of new water pipe in an urban street, I imagine they'd estimate excavation, shoring, bedding, placing pipe, testing the pipe's installation, backfilling and resurfacing before giving an answer to the nearest hundred dollars. If I inquired whether they thought the pipe would hit other utilities that would have to be relocated, they probably would admit that it's likely to happen since there are many utilities in most urban streets. However, they then might add they didn't include such utility relocation costs because I hadn't said they should consider that possibility. It is any such limited vision that exacerbates the difference between initial and final cost estimates.

Further, estimates often fail to reflect the sensitivity of the input. Consider that with final (100% complete) bid documents, the difference between the low and second low bidders still may be more than 5%. If with 100% complete data, the cost difference can be 5%, how accurately can we estimate the final cost during the conceptual stage when all we have are conceptual plans for something that isn't going to bid for say two years? We're lucky if an estimate at the conceptual stage is within 50% of the final amount given the usual changes between conceptual and final design, the unknown cost of materials two years in the future, the number of change orders that typically occur during construction, etc. To mitigate cost risk, I realized we needed a better system for preparing cost estimates.

From that realization, I began telling cost estimators to give me a range of what they think something will cost, with 95% certainty they think the range is reasonable. No more give me an estimate to the nearest hundred dollars, but set a range. As plans become more complete, the range would obviously narrow.[22-6] I proposed that we develop a cost estimating program to provide more realistic estimates; TxDOT liked the idea and increased our scope to develop an estimating program for them.

After attending a meeting in New York, I was flying back to Houston out of Newark Airport on Sept. 11, 2001, on an 8:00 am flight. As fate would have it for my fellow passengers and me, terrorists selected two other flights departing Newark Airport at 8:00 am to hijack. I was following our plane's progress on an automated flight map on a bulkhead screen, and I wondered why we shifted our flight path heading from south-west to south. Soon after, the pilot announced we were going to land in Atlanta, and he would tell us more then. All the passengers speculated on what was happening, but it wasn't until we landed that we found out about the hijackings, and that the World Trade Center buildings had fallen. I knew there were several Parsons Brinckerhoff employees working in the WTC and feared for their safety. Subsequently, I heard they all were well, although other engineers I knew did not survive.

Upon exiting the plane, the only thing airport personnel told us was no plane would depart Atlanta for an indeterminate number of days. I assumed it would be difficult to find a hotel or rental car near the airport and wasn't sure what to do. I called the Parsons Brinckerhoff Atlanta office, hoping to find someone to put me up for the night, but couldn't reach anyone as the office was in the process of an early closing because of the uncertainty everyone had that day. By a fluke, I spotted two other passengers from my flight who were fortunate enough to get one of the last rental cars at the airport and were driving back to New Jersey that day. They took me along as a third driver, and we got back to New Jersey at 5:00 am the following day. I stayed in New Jersey until flights to Houston resumed a week later.

I started with four Deputy Project Managers on the Katy project (one of whom was Jim Rozek). All had run their own projects before joining the Katy project, and each was jockeying to be the #1 deputy. The four were all highly competitive, and I had to keep them focused on helping TxDOT rather than clashing with each other. I said their objective should be to convince TxDOT of their own value if they wanted to succeed me, as it likely would be TxDOT who would decide my replacement. I then gave each a chance to shine before the client. For example, I rotated whom I appointed as Acting Project Manager whenever I was away from the office. I also alternated who ran internal and external meetings to give them further exposure. After that, the four worked together better — not perfectly, but better.

One deputy, Paul Weldon, was an officer in the naval reserves and after 9/11 was called to serve in Afghanistan. TxDOT was upset he was leaving the project, stating they preferred we maintain our original staffing, ignoring in this case it was not possible. They discounted my comment that we should honor Weldon's willingness to serve our country. I had expected to leave the Katy project by the second year, but stayed as Project Manager for five years because of TxDOT's position on our staffing. By then, our efforts were 50% complete, and three of the four deputies had moved onto other projects. That left one to replace me while I was looking for my next project management assignment. At that point, I became a part-time PIC on the Katy project. Four years later, I resumed the Project Manager role on the Katy as it wound down to completion. The Katy was a rare, large project for me, in that I was actively involved from commencement through completion.

My initial intent was to become a licensed Texas PE. However, I soon realized that because I never created a record with the National Council of Examiners for Engineers and Surveyors, it would be onerous for me to fill in Texas' numerous application forms, given the hundreds of projects I had worked on and that many references for my early efforts were deceased. Unfortunately, Texas and New York (where I had passed the test to get my initial license) didn't have a reciprocal agreement on licensing engineers that would have made applying for a Texas PE relatively easy. For those reasons, and assuming I would leave the project within two years, I opted not to apply for Texas licensure and had a licensed deputy sign engineering documents. If I had imagined I would remain involved with the project for eleven years, I certainly would have made the effort to become licensed.

CHAPTER 23

FIRM-WIDE DIRECTOR OF PROJECTS

I had been the Katy Project Manager for five years, when in 2005, my deputy took over for me, and I became the Katy PIC. While looking for my next project to manage, and as a temporary measure, I agreed to be Director of Projects for Parsons Brinckerhoff, although I previously hadn't enjoyed serving in a staff position.

As Director of Projects, I reviewed the firm's largest proposals and recommended pursuit of those I deemed suitable for the firm. I also reviewed performance of projects that warranted additional oversight and chaired the committee which created training programs for Project Managers and Project Administrators.

Over the years, Parsons Brinckerhoff developed a series of project management training modules, of which I prepared two modules ("Introduction to Project Management" and "Project Administration"), a major addition to the "Quality" module and portions of the "Negotiations" and "Risk Management" modules. For some modules, I created an exercise whereby trainees would work in teams to identify solutions to a problem. The intent was for the trainees to learn the value of brainstorming with others possessing different skills. Those five modules were fun to teach, as I had a number of war stories I could use to highlight key points. The more I did it, the more I found training others to be very rewarding.[23-1]

After serving a year as Director of Projects, the position was eliminated by new upper management in the domestic company, where I was based, after a company-wide reorganization. The new management seemed to have forgotten how important projects are to success. For example, a new domestic company's policy said discipline managers unilaterally would assign the discipline staff to projects. I asked if I, as the Project Manager, could reject staff assigned to my project I considered unacceptable. The reply was, "You can talk about it with the discipline leaders."
I said, "Okay, and if after talking I still want to reject staff, may I?"

"Everyone should collaborate," was the response.

My retort was, "If I deem personnel you want to assign to the project are unqualified and I can't reject them, then I'm not the Project Manager, you are."

My point was if Project Managers are responsible for the success or failure of projects, they have to have the right to approve who works on their project. Unfortunately, while I was well aware I was senior enough to challenge discipline managers, I knew most junior Project Managers probably could only grumble.

Because I felt project management was being given short shrift over general management, half in jest I said we should interview people to be Project Managers, and those who couldn't qualify as Project Managers should be considered for general management. Rather than causing the domestic company's leadership to rethink their approach, my facetious comment was ignored. An example of how sarcasm rarely works as one hopes it will.[23-2]

CHAPTER 24

EXTERNAL EFFORTS

In addition to my everyday job, I often was able to fit in a number of non-company efforts that helped me develop key skills and network with people useful in my career. Following are examples of what I did both to become a better Project Manager and to give back to my community and profession.

Community Service:

I wanted to perform community service and looked for ways to do something useful in Verona, New Jersey, the suburban town of 15,000 people where we were living in 1977. When I heard no one was running in the upcoming election for the two open seats on the five-member Board of Education, I filed papers to be a candidate. A New Jersey school board member is an unpaid state official, and the election process, while non-partisan, is very formal. I'm not sure I would have run had I known two other candidates eventually would file, and there would be three of us competing for two seats. Believing I had something to offer, I decided to stay in the race. That decision was fortunate, as I didn't realize how much campaigning would improve my management skills.

The other two candidates were educators, and I expected they would look to exploit their better understanding of educational issues against me. My approach was to change the playing field, so my managerial skills would be important in the voters' eyes. Because there already were two educators on the board, I sought to convince voters that a board member with business experience, and not more educational skills, was what the board needed — that I would bring business discipline to the board. While the board is non-partisan, the town's Republican and Democratic leaders pay close attention to the school board election, probably because the school budget is larger than the town's. Leaders from both parties contacted me to find out where I stood as I was a registered Independent and unknown to them. Once they realized I had no political agenda, each helped in my campaign.

EXTERNAL EFFORTS

To convince people to vote for me, I wrote press releases, prepared and gave speeches to local community groups, and participated in public debates. Similar with many engineers under 40, I had no experience writing anything for newspapers and was nervous talking in public. However, the more I wrote and spoke, the more skilled I became at each, enhancing my ability to be a better Project Manager. My campaign was successful, and I was one of the two candidates elected to the board. Even had I lost, running for the Board of Education was beneficial to me as a manager.

After being sworn onto the board, I had even more opportunities to improve management skills. As the only board member with business experience, I served on our negotiating teams with the school district's three unions, which upgraded my negotiating skills. I also negotiated the contract with the board's architectural consultant. Further, being on a board enabled me to appreciate how group dynamics work; I saw how some people at a meeting never seem to stay on topic, while others push their own agendas. To be a successful board member, I had to develop a style that enabled me to work for the common good, while influencing others to support my position on issues. Even when I felt I knew more about an issue than others on the board, I learned to respect everyone's opinion, as each board member was equal with one vote. My style evolved to become consensus building and non-confrontational.

The term of office was three years, and I was unopposed when I ran for a second term. The other board members elected me President each of my last four years. As President, I learned how to lead and structure discussions to reach satisfactory conclusions. I presided over about 50 public meetings, and whenever local citizens addressed the board, I'd listen patiently to their concerns and complaints and try to respond appropriately. However, when citizens grandstanded, I'd let them rant for the allowable five minutes. Then, rather than responding and keeping the fire going, I calmed the situation simply by saying, "Thank you for your comments."

When I announced I wasn't a candidate for a third term, both political parties tried to convince me to run for Town Council on their ticket. I declined as I felt it was more important to focus on my career at that time, which involved an increasing amount of travel. Summarizing, my Board of Education service helped me improve many of the skills I needed in advancing my career, including those related to leadership, negotiations and how to influence others. Further, serving my community gave me a sense of personal accomplishment. I heartily recommend everyone give something back to their community, as they will benefit in many ways.[24-1]

External Organizations:

Over the years, I was active in many outside organizations. I define "active" as doing more than paying dues and attending periodic meetings. Each organization had a different purpose, and collectively they helped me network, acquire new technical and managerial skills, identify new marketing opportunities, and give back to the profession.[24-2] It also was personally satisfying whenever I was recognized for efforts by my peers in those organizations.

The first organization in which I became active was the American Road & Transportation Builder's Association ("ARTBA"). ARTBA showed how our industry can influence governmental policy makers. I chaired the ARTBA Energy Council and its Planning & Design Division's Highway-Bridge Committee. Through my involvement, I developed contacts of long-lasting, personal benefit with principals of many firms typically competing against Parsons Brinckerhoff. I learned employees of those firms mostly are good people and not evil enemies to be hated.[24-3]

ASCE is civil engineering's leading professional organization and has been an excellent forum for me to serve my profession. I was a student member of ASCE in college and joined national ASCE after graduation, where I was assigned to the Metropolitan Section, the local section for the NYC area. In the mid-'90s, the Met Section was forming an International Committee and looking for candidates to be chair. Knowing I had much to offer based on my experiences in Hong Kong and Turkey, I applied for the position and was appointed chair. One of the International Committee's responsibilities was to host foreign engineers visiting NYC. To enhance the visits, I suggested we create a pocket guide of local civil engineering projects. The 36-page "A Guide to Civil Engineering Projects in and around New York City" took two years to come to fruition, but was an instant success. Over time, I was elected to the Section Board, chaired the Audit Committee, and elected Section President. I received the Thomas C. Kavanagh Award in recognition of service and dedication to the Section and profession in 2005.

While chairing the Section's International Committee, I met Mike Salgo, a former national Vice-President of ASCE and Met Section President. We bonded, and I discovered he had attended DeWitt Clinton High School, albeit twenty years before me. Salgo recommended me for appointment to national ASCE's International Activities Committee that coordinated the society's global efforts, and a few years later I became its chair. Subsequently, I was elected to the national Board of Direction in 2001, representing New York, New Jersey and Puerto Rico, and served on the national Finance Committee and Transportation Policy Committee. After my three-year term as director ended, I was appointed to such national committees as Audit, Professional Practice, and Licensure & Ethics and chaired the Task Committee to Study Professional Liability & Risks/Claims Management. In addition, I served on the Editorial Board of ASCE's "Leadership and Management in Engineering" journal.

I also was appointed as one of the three ASCE representatives on the Hoover Medal Board of Award. The medal is named for President Herbert Hoover, a mining engineer, and is awarded annually to an engineer for outstanding civic or humanitarian activities, constituting distinguished public service. Representatives from five engineering organizations (the American Society of Mechanical Engineers, the American Society of Civil Engineers, the American Institute of Chemical Engineers, the American Institute of Mining, Metallurgical and Petroleum Engineers, and the Institute of Electrical and Electronics Engineers) make up the board which is responsible for selecting the medal's recipients.

I always seek ways to show appreciation to CCNY for providing me with a free and comprehensive education. One way has been by chairing the college's Department of Civil Engineering Advisory Board, which affords an opportunity to influence the curriculum so it's appropriate and relevant to what graduates will face in the working world. Given the limited number of courses one takes in college, the curriculum never will provide everything needed to be a successful engineer. That said, it's my sense today's engineering curriculum at CCNY probably is more relevant to the practitioner than the curriculum when I was in college. At the time I wrote these memoirs, the Accreditation Board for Engineering and Technology (which accredits engineering programs nationwide) was proposing new student outcomes relating to project management. Students benefit when project management principles are incorporated in basic classwork. I had to learn about project management from mentors and on-the-job training.

Remembering how much I benefited from seminars held by visiting specialists when going for my Master's degree at Poly, I provide financial support for a seminar series at CCNY. Through the series, visiting engineers from academia, government and the private sector present topics of interest to civil engineering students and faculty.

Publications, Papers & Presentations:

The editor of McGraw-Hill's *Handbook for Civil Engineers* asked Dick Duttenhoeffer to completely update the chapter on "Highway Engineering" for the Second Edition of the handbook. The original chapter had been written by engineers from another firm. Duttenhoeffer asked Vik Kirkyla, a fellow Highway Engineer, and me to write the text and said he would review our efforts. We agreed and parceled out the chapter sub-sections between us. It was a far more intensive effort than I had imagined, but a great learning experience. The time spent in research to find the best information to include in the chapter served me well for many years, as it refreshed my understanding of several topics I hadn't worked on for a while. In addition, I gained instant credibility as a Highway Engineer in potential clients' eyes whenever I gave them an autographed copy of the handbook. A few years later, Aivars Delle joined the team when we updated the chapter for the Third Edition.

I wrote numerous papers and articles and gave many presentations. For example, Rozek and I co-authored a paper entitled "Program Management," presented at the Civil Engineering Conference in the Asia

EXTERNAL EFFORTS

Region in August 2004, that included some topics mentioned in this memoir. I guest lectured at the Middle East Technical University in Ankara, Turkey on the topic of engineering management and twice at the University of Houston, Texas, once on engineering management and once on drainage design. I also spoke at CCNY on "Project Management Challenges of Major International Engineering Projects." In my papers and presentations, I looked to avoid bragging about what I did, but rather described why I took the actions I did. I felt that approach gives readers and attendees the most benefit from my experiences.

I chaired "Public-Private Partnerships in Transportation," a conference of the California Engineering Foundation held in 1990. Among other responsibilities, I selected program topics and scheduled panelists and speakers, including Congressman Ron Packard and a future U.S. Secretary of Transportation (Elaine Chao). All in all, chairing that conference, while performing my then day job managing CTV, was possibly the most difficult non-project assignment I ever had. After that experience, I never criticized a conference chair.

Chapter 25

PROPOSAL MANAGEMENT

Managing a proposal (i.e., the tasks done to convince a potential client to select you to perform an assignment) is identical to managing a project, as each requires a scope, schedule and budget, albeit your firm is an internal client and there are no revenues to be earned in the proposal effort. A critical metric in proposal management is turning in a credible proposal on time. Perhaps bizarrely to some, I don't see winning an assignment as a primary metric, because the reasons why a potential client selects one firm over another often is out of a Proposal Manager's control. That said, when a Proposal Manager wins less than 15% of all proposals pursued, it's an indication of a possible problem, while winning over 30% is a positive indicator.

From 1989 to 2005, I managed or participated in many proposals, and the firm won most. Of course, I didn't win them by myself as others had major roles in each win. That said, I certainly believe my performance at those interviews where I was the proposed Project Manager had much to do with the success rate. I had discerned a winning formula at an interview was convincing the selection panel I would be better at taking the risk of project failure off their minds than our competitors. At that point in my career, I probably had managed more major projects than most anybody working for any firm and used that experience as a selling point. Namely, that I knew, far better than other Project Managers, what worked and wouldn't work.

Because of my record of success, many Area Managers sought me out to be their candidate Project Manager in proposals to potential clients. While it felt good to be desired, I avoided small projects where I would be expected to do much of the technical work (which others could do faster and more economically than I could). Also, in many cases winning a small project would have required me to relocate, which was especially disruptive to me and my family when the project had a short duration. For all those reasons, I focused on the larger, more complex projects where managerial skills were critical.

Following are case studies of three proposal efforts I was involved with, each with a different lesson to be learned.

New Jersey:
By the late 1970s, many clients had incorporated interviews in their consultant selection process, and I was a total failure, when I spoke at my first interview. The Parsons Brinckerhoff railroad group was pursuing a commuter-rail electrification project in New Jersey that required preparing a proposal and participating in an interview consisting of a 30-minute presentation and a 20-minute question and answer session. The railroad group's designated PIC (call him Dave) and designated Project Manager (call him Ed) were too busy to write the proposal, and because I had written several successful proposals, they asked me to be the Proposal Manager and prepare the electrification proposal for them, even though I wouldn't have a role on the project if we won.

I wasn't an expert in the technical details of rail electrification whereby a train uses electricity via a catenary instead of running on steam power. However, I did my best and wrote the technical scope with little help from Dave or Ed. I then asked Ed to read what I wrote so he could develop and give the 30-minute presentation at the interview. He said he would, but a few days before the interview, Ed said he didn't have time and I should give the entire presentation, while he and Dave sat nearby in support. By presenting, I violated a basic marketing rule, as I wasn't on the organization chart. Because I had minimal understanding of what I would talk about, I wrote out my presentation (something else we're told never to do) so Dave and Ed at least could review the content and correct obvious misstatements. As you no doubt guessed, they never reviewed my presentation. My reading text word-for-word for 30 minutes was boring for everyone and probably confused the selection panel who likely wondered why I was speaking instead of our Project Manager Ed. Not unexpectedly, the firm didn't win the assignment.

I was very upset at all the work I had to do when those who would have benefited the most if the firm won the project weren't willing to help in its pursuit. I suggest the next time anyone in Dave and Ed's position are that busy, they consider whether the best decision is to pass on pursuing a prospective assignment.

Oklahoma:
In 1996, I was asked to be Proposal Manager on the pursuit of a project in Oklahoma. The client was a regional toll authority, whose long-serving, national consultant was vulnerable to be replaced if rumors were true. A third competitor for the assignment was a small, local firm. I wondered if I brought added value to our team as the project wasn't a large one, and my experience managing large projects wouldn't be meaningful to the toll authority. Further, I wasn't an Oklahoman and had never worked on an Oklahoma project. In my opinion, Parsons Brinckerhoff, which only had a two-person office in Oklahoma, went into the selection process with no real game changer to put forward. Thus, the firm's only chance for success was if the client wanted a new national firm as their consultant.

The selection involved an interview conducted in public in the atrium of a government office building. We were the second team to be interviewed and decided it would be unethical to listen in on the first interview. We neither knew nor concerned ourselves if our competitors felt and acted the same way. Our interview went fine, nothing noteworthy happened, and ultimately the six-member selection panel gave two first place votes to each of the three teams. Unfortunately, Parsons Brinckerhoff lost the tiebreaker, and the small, local firm was chosen.

Oklahoma is an example of why large firms should think twice before pursuing small projects in new locales. As mentioned previously, always ask for a debriefing after a loss to learn what could be done next time to improve the likelihood of success. In some instances, you may discover which clients probably will never choose you over their favorite firms.[25-1] If that's the situation, you can either try to become a subconsultant to one of that client's favorite firms (part of a loaf is better than none) or decide to focus your efforts on pursuing other clients.

The Oklahoma pursuit was my only loss out of thirteen pursuits in a row that went to an interview; a very good record by any standard, but I still felt badly about the loss. Note, every so often a firm loses several pursuits in a row, which is very disheartening. For that reason, don't do what one Area Manager used to do which was to call every pursuit a "must-win." By telling his staff they had to win so many pursuits, he caused them to wonder if the firm was running out of work. Further, whenever they lost a must-win, it left staff fearful the office would close and everyone would lose their job.

Texas:

Parsons Brinckerhoff was pursuing the I-69 Program Management assignment in Texas, with Wally Dunn as the designated Project Manager and me as the PIC. I wasn't the Proposal Manager on this pursuit. The interview was held in a hotel meeting room in Austin, and each of the three competing teams was given space in the hotel to practice. While rehearsing, we noticed someone from one of our competitors lurking by our practice room, trying to spy on our run-through preparations, and I posted one of our employees outside the room to keep the eavesdropper away. In spite of the distraction, Dunn did a great job at the interview, and we were successful and won the assignment. Because the spy may have been acting on his own, we didn't report the incident to the client, although I did complain about it to a principal from the competitor firm. Lesson learned: Be aware that competitors may not act as ethically as they should.

CHAPTER 26

DIRECTOR ON AN INTERNATIONAL ENGINEERING COMPANY BOARD

In anticipation of Parsons Brinckerhoff Inc. board vacancies, Chairman Morris Levy surveyed the firm's largest shareholders asking whom they would prefer serving on the board. Keith Hawksworth (based in Hong Kong and President of Parsons Brinckerhoff International) and I headed the list. Then, when two vacancies opened on the board in 2004, Hawksworth and I became official nominees, and subsequently the employee-shareholders elected us to the seats.

As President of the international company, Hawksworth reported to the holding company President, who also was on the board. I was the only active Project Manager on the board, two levels below the domestic company's President to whom I reported. During my board tenure, there were two domestic company Presidents who didn't serve with me on the board. The first President treated me with respect and viewed my board service as a credit and benefit to him and the domestic company. The second, call him Ronald, acted as though he wanted me gone because he deserved that board seat. It was disappointing that Ronald never comprehended it was the shareholders who decided who held the seat, and it wasn't merely a reflection of one's operational title.

Chief Compliance Officer:

Everyone at Parsons Brinckerhoff felt it was a very ethical firm, yet one couldn't be blind to the fact that but one of its 12,000 employees doing the wrong thing could bring down the firm. To lessen that possibility, in early 2009 I was asked by the board to serve in a part-time role as the firm's first Chief Compliance Officer ("CCO").

As CCO, I was responsible to make sure the firm followed appropriate practices related to compliance with laws and regulations, such as the Foreign Corrupt Practices Act. Compliance with ethical standards also fell within the CCO's remit. One of my tasks was to heighten awareness of the whistle-blower hotline process, whereby employees could anonymously report any possible transgression they noted.[26-1]

Whenever I visited an office, I would ask the local manager to schedule time for me to speak to the staff about our whistle-blowing policy as a further means of spreading the word. Given that I was familiar with the pressures on those who did design, project management and operations, I looked to frame the discussion in ways that would resonate with all staff.

CHAPTER 27

REFLECTIONS

I know I'm most comfortable, and maybe you are also, with a standardized routine of do this first, then do that, etc., but I hope this book showed how project management isn't a cookbook effort. While many projects seem similar in approach, there often are significant differences in contract terms, local practices and people. For example, even when the client is the same project-to-project, your client counterpart may change over time as the original one moves to a new assignment, or even when your staff are the same, individuals become different as they gain more experience or suffer new family and personal pressures. And can you count on a cookbook approach to help you recover from the unexpected such as a project participant doing something unethical or having a financial crisis?

Dear Reader, the point is the Project Manager has to be ever vigilant on determining what systems are required to keep scope, schedule and cost in balance for each specific project and for each stage of a project. Many mistakes described in this book show how failure flows from complacency and over-reliance on what worked in the past.

Little did I expect that writing this memoir would clarify for me why I took many of the actions I did. It showed my success stemmed from an inclination to be a team player, a realization of the need to be organized, and a willingness to ask for help from others. Further, that I both learned from mistakes I made and established approaches to lessen the likelihood of repeating those mistakes. In sum, a disposition to learn and grow helped me advance further than I anticipated at graduation.

The memoir also showed the importance of being a problem solver who is able to envisage options to get to the desired endpoint, evaluate the advantages and disadvantages of different options, and confidently move forward with the option that made the most sense. Examples included creating a computer spreadsheet on I-787 to deal with large arrays of numbers, placing four countries' flags on my desk in Turkey to curb nationalistic rivalry, developing a strategy to advance metrication, and creating a program on the Katy Freeway to have more accurate cost estimates.

My career started in the design phase of engineering, but my real achievements occurred after I moved into engineering management and leadership. Recently, I've seen an executive temperament described as "organization, discipline, calm and restraint." While I had no knowledge of that definition during my career, I believe I evolved to fit that description as I emulated practices of people I respected. I choose the best practices from each and molded them into my own management style.

Early in my career, I occasionally would lose my temper or composure. At some point, I grasped doing so wasn't professional, nor did it set a good example for others. I stopped cursing (something I rarely had done before, anyway) and tried to convey I had everything under control — even when I wasn't fully certain things were yet under control. Also, having noticed how negative people dragged me down, I always tried to be upbeat and full of energy, so others would feel the same; I typically said a breezy, "Good morning," upon arriving to work.

My philosophy is to be happy and sleep peacefully knowing I did the best I could. Life's too serious not to see the more humorous side of things, and I often used self-deprecating humor to show we're all imperfect. For example, by joking when the wrong slide went up at the Guam interview, rather than panicking or blaming the technician running the slide show, I kept the interview panel on our side.

I chose a wonderful profession, one in which I was able to go all over the world and see the fruits of my efforts. I retired proud of my contributions to the profession, and thankful that I'd had the opportunity to work with many exceptional people in my career. Further, I'm pleased the profession is well positioned going forward and delighted to see those now entering it are more socially conscious than my generation was at a comparable age. This new ethos is evidenced by the service many recent graduates give to organizations such as Engineers Without Borders and Bridges to Prosperity and their commitment to sustainability, diversity and ethics.

If I were advising someone starting out in their career, I would say do what you find enjoyable and not solely what you think you need to do to be financially successful. No one gave me that advice when I started, although I ultimately found that path myself. When you enjoy what you're doing, you have a positive attitude as you go to work, there's less stress on you, and you're a more pleasant person when interacting with co-workers.

Keys for Success as a Project Manager:

Project Managers must maintain discipline to achieve success, given they are fully responsible for the success or failure of a project in the system of many organizations. Yet typically, project resources belong to someone else, and most key decisions (e.g., which clients to pursue, which major firms to team with, and what reimbursement terms to accept) are made by someone more senior than the Project Manager. Regrettably, too many in corporate management don't understand what a Project Manager has to do to be successful and, therefore, have no empathy for a Project Manager's difficulties. I'm reminded of the comic strip, "Wizard of Id," where a farmer is petitioning for redress from the king. The farmer claims the king's troops knocked down all the fences on his farm so his herds ran away, then the troops trampled all the crops in his fields, and next they ate all the food stored in the barn. In the last panel, the king says, "No one said farming was easy."

Well, project management isn't easy either, but I still found it exhilarating to confront its challenges.

There are many ways to become successful, but following are some keys to the reason I achieved what I did:

1. Work harder than others/Become essential/Be proactive[27-1]: It was propitious that I worked for and learned from Jenny and Lammie, two managers who never coasted and always put in the extra effort. Their perseverance set the tone, and they were great examples for me to follow. I also tried to be indispensable to my supervisors, and someone who always understood how my current assignment fit into the big picture.

Comic strip Dilbert's supervisor showed he understood this example when a storm was approaching, and he goes on the office intercom and announces that all non-essential employees could go home. He then looks out the window at those leaving and says to himself that determining who should be in the next lay-off will be easy.

2. <u>Study & learn from your mistakes and those of others</u>: Bad things happen to good people, and we must continually learn from the mistakes of others or regret the consequences. Each mistake gives us something of value to study, providing we're willing to learn. True risk management involves identifying what went wrong in the past and then determining what systems have to be put in place to lessen the likelihood it will occur again, so we don't repeat a mistake previously made. It's a matter of constant self-criticism followed by self-improvement. We're better when we learn from the experiences we've encountered.

Carefully consider the taking of shortcuts, as each increases the likelihood something bad will happen. We know the right thing to do, yet repeatedly look for shortcuts thinking we'll save time or money. In most cases, we're lowering quality and increasing risk. As an example, don't believe you're saving time if you decide not to document minor items.

3. <u>Earn loyalty</u>: Lammie ran a survey where he asked people what their staff wanted of them and what they wanted from their supervisor. Most said they were a good supervisor because they were technically competent and that's what their staff expected of them. Next, they responded "honesty and fairness" is what they wanted of their own supervisors without realizing the dichotomy in their two answers.

The survey results helped me understand how important it is to be fair with everyone. I frequently asked those working with me, "What can I do to help you?" which I hoped showed I was willing to help. In addition, I looked to collaborate with others as a sign I respected their efforts. If you want to become the type of leader for whom you would like to work, you must transcend egotistical needs by showing those you work with that you believe what's good for the organization is more important than what's good for you personally.[27-2] Additionally, one of Lammie's sayings that stayed with me was to take care of my client and my people, and the profits will follow.

4. <u>Be flexible</u>: The setting I worked in kept changing, and I continuously had to adapt to be effective and do my job properly. No doubt, the future will have its own share of major changes in how engineering is accomplished, requiring each new generation of Project Managers to adjust and accept those changes with flexibility and a positive attitude. Following is a brief summary of the conditions I faced by decade:

   a. <u>First Decade</u>: The three firms I designed highways for (DeLeuw, Cather & Brill; Brill Engineering; and Parsons Brinckerhoff) were all relatively small, engineering consultant partnerships. Most employees were based in Manhattan, working on the design of highway, tunnel and bridge projects. Mainframe computers were just becoming an engineering tool.

   b. <u>Second Decade</u>: The number of Parsons Brinckerhoff owners (partners) doubled in size as the firm opened up major offices in San Francisco and Atlanta and became a leading transit firm. I began managing projects.

   c. <u>Third Decade</u>: Although former partners controlled most of the shares, Parsons Brinckerhoff started operating as a corporation, with about a hundred shareholders. The firm had offices in some twenty cities in the US, and there was a major office in Hong Kong providing building M&E services. The firm no longer had a separate department of project managers. Computer aided drafting had begun. I managed my first mega-project.

d. <u>Fourth Decade</u>: Parsons Brinckerhoff was operating as a true corporation with over a thousand employees owning shares. It had doubled the number of its US offices and had several international offices. The firm was a major international presence in power. Desk-top computing was becoming the norm. I worked in Turkey, California, Hong Kong, Budapest and Detroit on various project types including design-build and P3.

e. <u>Fifth Decade</u>: Five thousand employees owned shares in Parsons Brinckerhoff, with no single shareholder owning more than 2% of the outstanding shares. A majority of the Board of Directors were non-employees. Half the 12,000 employees were based outside the US, and the firm truly was an international firm. Computer aided design and cell phones were ubiquitous. I worked in Buffalo, Princeton and Houston.

f. <u>Sixth Decade</u>: Balfour Beatty, an international contractor trading on the London Stock Exchange, bought Parsons Brinckerhoff, which kept its name and operated as a subsidiary company performing engineering services.[24] I worked in Panama and Guam, before retiring from Parsons Brinckerhoff and then serving as a board member of a private, California toll road.

You need to be flexible as you encounter differences from project to project: different types of project, clients, and staff, and from projects requiring relocation. Following are tips that I found worked in various situations:

- Project type:

---

[24] Three years after I retired, WSP, a management and consulting firm, bought Parsons Brinckerhoff from Balfour Beatty. In 2017, Parsons Brinckerhoff was renamed WSP.

i. Surround yourself with experts in disciplines that aren't your strength.⁽²⁷⁻³⁾ Understand that majoring in project management may result in the inability to maintain expertise in your primary discipline, and that one day you may have to add an expert in your own discipline to the team.
ii. Increase the number of deputies and key administrative staff as a project's size increases.⁽²⁷⁻⁴⁾

- Client:
    i. Understand that a private sector client (e.g., a contractor or developer) is likely to be more concerned with meeting schedules and beating the budget than a public agency client. Such a situation puts pressure on quality maintenance, so sound quality checking protocols must be instituted.
    ii. Make the effort to learn a new client's procedures and how fairly they will treat you.⁽²⁷⁻⁵⁾
    iii. When your client counterpart changes, expect the new one will change some decisions the previous one made.⁽²⁷⁻⁶⁾ Protect yourself from costly changes and criticism by contemporaneously documenting in writing all directives from your current counterpart.

- Staff:
    i. You are less likely to be disappointed in the performance of staff who are new to you, when you know their strengths and weaknesses at the start.⁽²⁷⁻⁷⁾ Vet key staff before they are assigned to your project.⁽²⁷⁻⁸⁾
    ii. Allow more time for do-overs when using inexperienced staff. Anticipate that inexperienced staff won't perform to the same level as experienced.⁽²⁷⁻⁹⁾
    iii. The more staff are scattered in different locations, the more difficult they are to control. Make sure

you have sufficient additional budget and duration to be successful.[27-10]
    iv. Start developing your successor from Day 1.[27-11]
- Relocation:
    i. Relocation puts pressure on the work-life balance of you and your family. Allow for the extra stress you will encounter. Don't assume a relocation will solve an existing family dysfunction.
    ii. Author James Michener said, "If you reject the food, ignore the customs, fear the religion and avoid the people, you might better stay home." Quickly acclimate to wherever you're relocating. Embrace where you are and invest in the same issues as the locals, dress like the locals, eat where they do, and support local sports teams, charities and cultural institutions. Engendering shared goals with local co-workers makes a stay more pleasurable. Of course, all this starts with a willingness to consider relocation for your next assignment.[27-12]
    iii. Make the effort to learn what skills the locals have, rather than prejudging the local's talents as compared with your own.[27-13]

Forks in the Road:

I faced many decisions where the path chosen would have a significant influence on my life and career. The following six were particularly testing to make:

1. <u>Public Sector vs. Private Sector</u>: My first major decision was whether to work in the public or private sector. While I think I would have done well in the public sector had that been my choice, the private sector gave me the chance to work on a variety of projects. These projects involved different disciplines ranging from planning through implementation; different

markets such as highway, power and buildings; and different locations both domestically and internationally.

2. Stay at Parsons Brinckerhoff or Leave: Deciding to stay at Parsons Brinckerhoff rather than moving every two years was a complicated choice early in my career. Once I realized the growth potential was good and how much I liked my co-workers, spending 50 years at Parsons Brinckerhoff was both rewarding and fulfilling. When the firm became international, it gave me the chance to see parts of the world I might not otherwise have visited, such as Lima, Panama, Ankara, Dubai, Beijing, Hong Kong, Bangkok, and Sydney.

3. Decide Whether or Not to Provide Community/Public Service: Another key choice I made was to provide public service by running for election to the Verona, NJ Board of Education. Adding public service to my day job responsibilities and family obligations could have resulted in my failing to satisfy all three commitments. However, serving my community not only gave me personal satisfaction, but helped my career as I improved management, leadership, communications, and negotiating skills.

4. Ask for Help or Go It Alone: When I tried all the tactics I'd used successfully in the past and still everything seemed to be going wrong on Westway, I could have blamed the client or everyone around me. Instead, realizing I was the problem, I sought help before things got worse and my supervisor had to step in and replace me. With the help, I was able to turn things around, which set my career in motion as a successful manager of major, complex projects.

5. Operations Management vs. Project Management: The decision to eschew operations management and go back to project management enabled me to return to what I enjoyed doing.

6. Relocate or Stay Put: Agreeing to relocate domestically and internationally, meant being willing to work outside my comfort zone. Relocation eventually increased my avenues for

advancement. Further, my willingness to work overseas opened the path to serve my profession, when it led to the opportunity to chair the ASCE Met Section's International Committee and, eventually, to my election to national ASCE's Board of Direction.

Making a hard choice isn't easy, but being willing to do it is fundamental. As someone prone to caution and levelheadedness, I went against my normal aversion to risk taking and made those six decisions, with no guarantee of outcome. Fortunately, they turned out to be largely successful. Of course, had my foresight matched my hindsight, I would have done many things differently, although I doubt my professional life could have been much better than what actually happened.

High Points:

There were numerous highs in my career that carried me over the periodic setbacks. The most significant highs were becoming a licensed Professional Engineer, being chosen Westway's Project Manager, being selected Regional Manager, leading the International Road Federation delegation at a United Nations meeting in Bangkok, being elected to the boards of ASCE, Parsons Brinckerhoff and SBX, and receiving the CCNY Engineering School Alumni's Career Achievement Award. I'm more than satisfied that, collectively, these accomplishments validate I made a meaningful mark in my body of work.

The pride of winning 16 out of 21 pursuits that went to an interview (including ten in a row) was a high spread over two decades. It meant a lot that my firm had faith enough in me to propose me as their Project Manager, and that clients believed I could deliver a successful project for them. Each win was worthy of its own celebration.

Significant Projects:

Successfully managing one or two projects might make us think we're a project management expert. I say, "Not true," simply because we haven't yet made enough mistakes. It took me decades, both managing dozens of successful and some not so successful projects and making and witnessing numerous errors, to become capable to manage the different projects and situations I encountered, including:

- Varying project sizes (from small and simple to multi-disciplinary, mega and complex),
- Public sector, private sector and internal clients,
- Different delivery methods including design-bid-build, design-build, and public-private partnerships,
- Both domestic and international venues,
- Evolving technologies (e.g., slide rules to main frames to hand-held computers to computer aided design).

Russ Fuhrman, my last supervisor at Parsons Brinckerhoff, asked which was my favorite project. Thinking how to respond, I realized I couldn't limit myself to only one favorite project. Rather, I told Fuhrman of four projects that stood out as both valuable for my career and personally fulfilling. Because of changed events, I added SR-125 as a fifth project as I wrote these memoirs. My five favorites are:

1. I-787/Mall Arterial: The project where I developed basic project management skills and had a very collegial relationship with the client.

2. Westway: My first program management assignment where I learned how to be successful on large, complex projects and programs.

3. Ankara-Gerede Motorway: While the most difficult and challenging project I ever managed, it was where I learned what works in an international setting and on design-build projects with a contractor as the client.

4. Katy Freeway: An award-winning program management assignment where almost everything went perfectly, as I applied approaches I had developed over five decades. A rare, large project in that I was involved from its start to finish. Another project where the client relationship was very collegial.

5. <u>SR 125</u>: I added SR 125 (my first P3 project) to my final list upon retiring from Parsons Brinckerhoff, because it brought me closure to serve on the board for a new consortium of owners, after having resigned 20 years earlier as President/Project Manager for the original ownership consortium.

Epilog:

Reviewing the final draft of this memoir, I realized how much I would have benefited having something similar early in my career. It would have complemented the books and articles I archived over the years and helped me avoid mistakes I made. Certainly, it could have smoothed the transition from relatively easy engineering to complex management. Much of the science of project management is a list of do's and don'ts one has to be rigorous about following. The art of project management is knowing what to do in the grey areas. Less easily described than do's and don'ts, that knowledge often can be gained only through the experience of making mistakes and observing the mistakes of those around you. Fail to do what you should, and you've taken a poorly thought-out shortcut that you'll regret sooner or later. The numerous missteps mentioned in this memoir reinforce that point.

Senior management has an obligation to expand the skills of those engineers who find engineering easy, so they can become better managers and leaders. With that in mind, my hope is the numerous cues listed in Appendix A will help others avoid missteps in their career. I also prepared the summary of Good Practices/Lessons Learned codified in Appendix B as a way for me to pass on to others what I've learned from my successes and failures.

The opportunity to work with thousands of talented colleagues on exciting projects made my career special. While I noted some of those colleagues in the previous chapters, I'd be thoughtless if I didn't highlight Art Jenny and Jim Lammie, for special recognition in this chapter. Jenny prompted me to remember how important project management was each time he said, "The engineering is easy," in an ironic tone. And, it was because of Lammie reminding me to take care of my employees (which complemented advice from my wife, Mary), that I tried to be fair, honest and faithful with those I worked.

While I don't expect to practice engineering again, I enjoy attending the occasional technical seminar. Also, my on-going volunteer activities for CCNY and ASCE still bring me the satisfaction of giving back to my profession. I hope my efforts encourage others to serve their profession.

The combination of doing good works with integrity, serving my community and giving back to my profession enabled me to attain the pride I sought by studying civil engineering. I also found it very satisfying mentoring and developing those who will follow me, whether it was helping educate children in my town or students in my college, developing civil engineers as an ASCE volunteer, or training a co-worker. It was especially gratifying when I saw someone understood why I did something a certain way. Thus, my legacy, such as it is, is justified not just by the projects I did, but it includes the good works of those whom I helped train and develop. And isn't that what sustainability's really all about?

APPENDIX A

PROJECT MANAGEMENT ONE-LINERS

Based on experience, I've learned what will and won't work in most project situations. Instinctively, "one-liners" often popped into my head when I saw someone straying from what I thought was the best path (or even sensed I was the one about to do something wrong). These one-liners became cues for me to suggest a different approach when speaking with Project Managers about their projects. A characterization for most one-liners is they are a truism or maxim representing good practices that were successful for me. Many one-liners aren't something I created, but something I heard and tucked away in my memory for future use.

One-liners in this appendix are codified into four general categories:

1. **Style**: Things to consider when deciding which management style works best for you.
2. **Staff**: Things to consider when dealing with staff issues.
3. **Risk Management**: Things to consider when dealing with the risks projects encounter.
4. **Organization**: Things to consider when creating an organization structure.

1. STYLE

a. Good Project Managers learn from their mistakes; excellent Project Managers also learn from the mistakes of others. *(Experience shouldn't mean making the same mistakes, but with more finesse.)*

> Before starting a new effort, ask others what they encountered when they worked on a similar effort. Specifically ask about what they would do differently and why. Then, learn from their mistakes so you can avoid repeating those mistakes. Note, I've probably made every mistake there is; I just try not to repeat them.

b. Don't give him anything good to hit, but don't walk him.

The statement paraphrases advice from a worthless pitching coach to a pitcher. The coach is trying to guarantee that no matter what the pitcher does, the coach can't be faulted for giving poor instructions. Don't be that coach, but review all your directions to confirm they give realistic and useful advice and instructions to others.

c. Cost + Schedule + Quality = A Constant.

Telling someone to cut the budget and accelerate the effort without cutting quality is similar to the pitching coach's advice. While it can be accomplished, it's unlikely unless there is a major change in some project element. It's usually wishful thinking to expect something to be cheap, fast and good.

d. Door closed/door open – it doesn't matter.

If you sit in an office with the door closed, your team will think you're doing secretive things you don't want them to know. If the door's open, they'll think you're spying on them. Net-net: Talk frequently with your team so they know who you really are and are comfortable with whichever way your door is that day.

e. Keep the monkey off your back.

The monkey refers to a task you willingly accept for yourself and comes from an article entitled "Who's Got the Monkey" in the *Harvard Business Review* by William Oncken, Jr. and Donald L. Wass. The message is that you shouldn't let subordinates assign their work to you; don't assume action items from those working on your projects. I often said to my staff, "I give stress, I don't take it."

However, whenever someone tries to give you their problem, you still should start with, **"What can I do to help you?"** If they ask your solution to a problem, first ask them to describe the options they evaluated and the one they prefer. Only then, give them advice, always remembering they should leave the meeting with the action item and not you.

At the same time, I would tell my staff that they're authorized and expected to act if I don't respond timely to their questions, and that I will support their action and not second-guess them.

f. Size matters.

Complexity is proportional to size; greater size means greater complexity. Even if you're a great Project Manager on small and medium projects, it's unlikely you'll be great on your first large project. It's no different than a great designer of simple one-span structures is unlikely to be great on their first cable-stayed bridge. In both cases, continuous support and training are required to make the transition from small to large. Fail to ask for such support, and you're likely to be disappointed in the result.

g. If your only tool is a hammer, you see everything as a nail (*thank you, Maslow*).

People gravitate to whatever makes them feel comfortable. Conversely, they don't like to work outside their comfort zone. In addition, people do what's in their best interest. Likewise, accept that it's your obligation to make sure your approaches are justified and not just approaches with which you're most comfortable.

h. Building a cathedral.

The old story goes like this:
> Separately, I approached three construction workers doing the same task and asked what they were doing. The first said, "Laying stones," the second said, "Building a wall," and the third said, "Building a cathedral."

Now, which one do you guess is the most motivated?

Schedule monthly staff meetings of all staff (top to bottom) so everyone sees how their effort is an important part of a major undertaking — the project. As part of a good participatory management style, go around the table and have everyone talk about what they're doing.

i. Give credit; take blame.

When something on your project goes wrong, don't blame your staff. After all, if your staff wasn't up to the job, why did you give them the job to do in the first place? On the other hand, always give your staff the credit when things go well. It's basic Team Building 101. People go the extra mile for you when they see you both care for and respect them. As President Eisenhower said, "Leadership consists of nothing but taking responsibility for everything that goes wrong and giving your subordinates credit for everything that goes well."

j. Don't count on volunteers.

It's always great when someone volunteers to help on your project. However, you still must make people commit to a work effort via a scope, schedule and budget. Otherwise, you have no leg to stand on should volunteers miss targets. In fact, you'll be reluctant to criticize them because they were so kind to volunteer. Your task is to turn casual volunteers into active participants.

k. The only reason I'm paranoid is people are out to get me.

Let me paraphrase a routine the late comedian Danny Thomas told maybe 60 years ago, long before cell phones existed.

> It's a rainy night on a deserted country road, and a driver changing a flat tire finds his jack isn't in the trunk. Realizing it could be morning before another car comes along, he decides to walk two miles back to a service station he remembered passing. As the driver walks through the cold rain, he thinks the following to himself:
>
>> I'll ask the attendant to drive me back to my car with a jack so I can put on the spare. But, the attendant will probably be alone, and I bet he won't want to leave the station.
>> Okay, I'll borrow the jack, walk back through the rain, put on the spare and drive back to the station to return the jack.

But I bet he won't trust me to return the jack. He'll probably ask for a deposit to lend me the jack. How much could he want as a deposit? $20? Okay, I'll leave $20.

No, the jack may cost more so I bet he'll want $40 as a deposit. Okay, I'll leave $40.

But wait, he's a businessman. I'm sure he wants to make a profit. How much will he charge me to rent a jack for an hour? Maybe he'll want $20. Okay, I'll leave a $40 deposit plus pay him $20 to use the jack.

But he knows he has me over a barrel. I bet he'll ask for a $40 deposit and a $50 use fee. No, he'll ask for a $40 deposit and a $100 use fee. No, a $150 use fee.

And as he walks through the rain, the driver gets madder and madder as the cost gets higher and higher. Finally, the wet and weary driver arrives at the service station, flings open the door, and yells to the attendant, "Keep your lousy jack!"

The point of the story is don't start by assuming negative motives in people you never met or even those who disappointed or criticized you in the past. After all, most people want to do the right thing and are not out to get you.

l. Theory of Limited Objectives.

The theory is based on making sure you accomplish the highest priority objective first. Don't focus all your efforts on the second highest priority goal until the highest priority objective has been met. Of course, you can have multiple objectives; it's only making sure you have your priorities straight.

m. When you've earned the eggs for breakfast, also bring home the bacon.

This statement says you can't afford to rest on your laurels. When you accomplish a small success, look around to see what you can do to turn a small success into a major one. When your client (or supervisor) is pleased with what you're doing, it's a good time to identify other areas where you can offer to help the client.

n. Do the right thing. Would you want to see it on the front page of tomorrow's newspaper?

If you think something may not be the right thing to do, it probably isn't. Don't ask people to do something both they and you know is wrong. Most people want to do what's ethical and proper and, when asked to violate basic principles, eventually will quit their job and/or "blow-the-whistle" as they should. You'll sleep much better when you only ask others to do the right thing.

o. Perception is reality.

It's a hard, long journey to convince people (and clients) that their perception is wrong. Make sure it's worth the effort before starting the journey.

p. Sarcasm doesn't work.

Use sarcasm to amuse yourself; just don't say it or send it to others.

q. Don't start an e-mail war.

Talk face-to-face (or by phone if in different offices) and get to the important issue quickly. Be bigger than the other party and seek a compromise that both of you can live with.

r. Criticize the action, not the person.

When something goes wrong, don't say the people who carried out the approach were "stupid" or "foolish." People become defensive when they're personally criticized. Instead, say you're unhappy with the results or say you question the process or approach followed. When you say the approach wasn't a good one, you then can say why you feel another one would have been better. That style is more likely to get people to listen to you, rather than automatically discount your ideas.

In any case, when something goes wrong, your first reaction should not be to cry over spilt milk. Instead, focus on what has to be done going forward to solve the problem. I've seen a case where the contractor finds something in the field doesn't fit, and the resident engineer asks for a quick fix so as not to delay the contractor any more than necessary. What happens too often is those asked for the new design waste time trying to find out who may have made an error instead of trying to solve the problem.

s. Avoid shotgun memos.

A shotgun memo is written when someone does something wrong, and a memo or e-mail is sent asking everyone to stop doing whatever it is that was wrong. In those cases, don't send the memo to everyone, but just to those who transgressed. Otherwise, you demoralize all innocent staff because of your unwillingness to confront the individuals who did something wrong.

t. Can I borrow your watch?

Don't ask your client or supervisor what to do without recommending an option. For some, the definition of a consultant is one who asks to borrow your watch when you ask them what time it is. Provide value to your client or supervisor by providing recommendations.

u. Make your client counterpart a hero to their boss.

A letter or e-mail to your client counterpart implying blame to him or her for any reason (e.g., why you're overrunning a task's cost and schedule) won't make your counterpart look good in their supervisor's eyes. Word each piece of correspondence so it is something they want to endorse and pass upwards in their organization.

v. Avoid dealing more than two levels down.

If you constantly have to go more than two levels down in your organization to assess what's happening, question either your management style or the capability of those one and two levels below you. Something's amiss somewhere.

w. Look in the mirror.

A Project Manager is responsible for everything. If things are going wrong, start by reviewing your decisions and actions. Often you will find the corrective action is yours.

2. STAFF

a. No one's perfect, except the person your spouse could have married.

Find out the strengths and weaknesses of those working with you, so you can play to their strengths and compensate for their weaknesses.

If someone wants to assign a current employee to your project, ask to see that person's last two performance evaluations. Such evaluations aren't always relevant, but when you learn an individual's strengths or weaknesses from the evaluation, you have a better idea what tasks to assign them. If you accept unqualified staff, don't blame them when they fail to perform.

Everyone will fail at something eventually. Don't come down too hard on someone who usually succeeds, but has an occasional failure. At the same time, don't fail to perform oversight of everyone, so you can spot a trend in the wrong direction before an error becomes a disaster.

Assume half your staff and half of all subconsultants will disappoint you. Since you don't know which half, keep your eye on all of them until you know.

b. Don't assume others are as good as you.

When preparing a budget, assume average competencies, not those of the best staff, as you never know who will actually do the work.

Cut rookies some slack until they've had a chance to gain enough experience to do the job as well as you. Until then, provide continuous support and training to help them make the transition from rookie to veteran. Fail to provide support, and you're likely to be disappointed in the result.

c. It's easy when you stand on the shoulders of good people.

Always strive to have the best people around you. Half the people in any organization are below average, and there's little reason to be the one struggling with the poorer performers. Hire and recruit above-average performers and reject anyone offered that doesn't meet your high standards. Ease up only when satisfied you have a sufficient number of the right staff to do the work.

Interview and approve the staff a subconsultant wants to assign to your project. Make sure they give you their best staff in whatever discipline you're asking them to perform.

d. When people are headed in the wrong direction, for goodness sakes don't motivate them.

Say someone is poor at marketing, but does well at design. What makes the most sense: To assign them marketing tasks or design ones?

Assigning people to logical tasks doesn't mean you don't give them opportunities to try other things, and, maybe over time, become skillful in those areas. The key thing is you don't want to pin your success on the **wish** they become skillful. Their becoming skillful should be an added bonus, not a job requirement.

Following are two ways to help technical staff grow into management roles:

(1) When you'll be away for a few days, consider rotating the acting Project Manager position among key staff. That approach helps you decide which one deserves the ultimate designation as your successor. It also gives more individuals the chance to impress others about their ability to cover in your absence.

(2) Assuming you have (and you should) a formal weekly meeting with your client, designate different key staff to lead the meeting for a few months at a time. Have the person truly run the meeting in all aspects: prepare the agenda in advance of the meeting, go the through the agenda at the meeting, and approve the draft minutes for distribution after the meeting, while you participate as one of the regular attendees.

e. It's easier to struggle with a sick jackass than to carry the wood yourself.

Don't automatically complain about those who are below-average performers as there's always a task for them to do. Have them do something that frees you up for the more important tasks.

3. RISK MANAGEMENT

a. A shortcut is another way of saying I'm increasing risk.

We know the right thing to do, yet we continually look for shortcuts thinking that we'll save time or money. In reality, we're lowering quality and increasing risk. We understand we shouldn't drive too closely to the car in front of us, yet many of us sometimes do, even though we realize doing so means we're more likely to have an accident. The same principle applies in management: Take a shortcut, and you increase the likelihood something bad will happen.

For example, we recognize everything sent out must be checked. However, when you're running late on a transmittal, do you decide to take a shortcut and send it out unchecked? Because no one's perfect, this could be the occasion where there will be an error in the work. Is it really worth the risk to send the work out unchecked? Avoid the shortcut and check the work!

Remember: **Do it right the first time.**

b. The kids are quiet.

Do you really think everything's okay if your children and their friends have been upstairs for hours without making a sound? Don't you think you better check on them? The same concern applies to anyone working on your project who has been quiet for too long.

c. Hope for the best, but plan for the worst.

It's foolish to believe past success means you don't have to develop contingency plans. Maybe it will be something out of our control that jumps up and bites us or maybe something we should have foreseen. In any case, failure to be ready for the worst-case scenario is a common occurrence. When we're used to things going perfectly, we often forget that something is likely to go wrong. Murphy's Law has not been repealed. And Murphy was an optimist!

What's going wrong in one office or on one project is going wrong elsewhere and eventually is likely to go wrong in your office or on your project. Be ready to stop it from becoming a problem and then a disaster by pre-planning to prevent its occurrence in the first place.

d. The five Ps: Prior Planning Prevents Poor Performance.

Avoid rushing into the first step of whatever you're doing, but rather spend the time to plan it out first. If you always try to do things on-the-fly, you won't be as prepared as you should be to deal with whatever happens to go wrong.

e. The engineering is easy.

Never underestimate the amount of effort required to do the administration and management on a project. A Project Manager is more than just the Project Engineer. A Project Manager has responsibility for the administration and management as well.

For example, project start-up is more complex than many realize. The client wants you solving technical issues on Day 1, while company management wants you to have a Project Management Plan, Quality Control Plan, Work Breakdown Structure, etc. in place the same day. To serve both those masters, arrange for extra management and administrative support well before notice-to-proceed to help do the early paperwork required at start-up.

f. The 30-Minute Rule.

Everyone has 30 minutes to tell their supervisor and client counterpart when something goes wrong or significant happens. You don't have to know how to solve the problem at that point, but you honestly can say you've started the process to find a solution. Remember, you don't want to receive a call from your client that they received a call from their supervisor about a problem your counterpart knew nothing about.

g. When I learn about something you did wrong, it better be from you.

You want people to admit their errors to you, before you learn about an error from another source, such as your client or the media. Tell people you'll share responsibility with them for solving a problem when they tell you about it first; otherwise, they take full responsibility for solving it.

h. Triple the time and cost given to you for software development.

Software development rarely seems to finish on time or on budget, often because end-users are not able to define what it is they want well enough to get what they need. Because I found initial estimates of time or money are typically inadequate to finish developing software on schedule, I often doubled or tripled time and cost estimates to have sufficient contingencies.

i. Sooner or later financing becomes an issue on a long-range program.

Even when a project starts with the owner saying we'll have all the required money in place when needed, there always seems to be a financial set-back at some point.

j. When five people owe you something next Friday, it's more likely none will deliver on time than all five will deliver.

When delivery by a set date is important to you, you must start oversight early to make sure all five know how important it is to deliver on time. Keep checking to confirm everyone is working on your task. If you don't, it's almost a guarantee at least one of the five won't make your deadline because they're struggling over some issue or even have shifted onto someone else's project. At the halfway point, ask to see their work-in-progress. You learn there's a problem when they can't show you progress. **You get what you inspect, not what you expect.**

PROJECT MANAGEMENT ONE-LINERS

k. Govern by schedule, and costs usually come within budget.

Original budgets typically are based on completing the work on schedule. For that reason, keep pressure on staying on schedule, because it's likely the budget will be exceeded if people don't finish on time.

It ain't over 'til it's over. Don't tell anyone something's finished if you haven't delivered it in final, checked form to them with your recommendations.

l. A task force always appears to be on budget and schedule.

A task force is a set number of people working on a specific work effort; for example, a task force of ten full-time people for ten weeks (10 x 10) results in 100 weeks of budgeted effort. At the end of three weeks, the 10-person task force will have expended 10 x 3 = 30 weeks of effort. However, the effort expended has no relationship to progress. You have no way of knowing if the task force's progress is 30% complete; you only know that the cost expended is 30% of the budget. To know if you are ahead, on or behind budget, you must assess both the effort performed and the effort remaining to be performed.

m. If it isn't in writing, it doesn't exist.

Too often, parties disagree on what was decided at a prior meeting. You think you assigned a task and they don't. You think a conclusion was reached, but they don't. Without timely documentation such as minutes, there's no proof something happened nor is there an opportunity for a party to contest someone else's understanding of what happened. By timely minutes, I mean within two weeks' maximum, but no later than the next time the group convenes. And the best minutes are prepared and distributed straightaway after the meeting ends when facts and nuances still are fresh in everyone's mind.

n. E-mails shouldn't count.

E-mails can last forever, yet many are prepared too casually and thus fail to adequately convey meaning and intent. Also, e-mails often are not placed in an easily retrievable document control system, which means it can be difficult to create a researchable record. Therefore, I required on my projects that all information is transmitted and received via hard copy letters of transmittals or transmittal forms. My approach meant an e-mail was not considered a formal project directive and no individual should have to act on the e-mail's contents.

Note even though I said e-mails didn't count on my project, I knew that probably wouldn't apply in a legal proceeding. I merely was trying to make sure writers were careful when creating messages.

o. Read your contract.

In fact, read it at least once a month from cover to cover, including every attachment, appendix and reference document. You must know its terms better than anyone else. It's a clear example of where knowledge is power.

p. How much will you pay when you're choking?

The story goes like this:
> A man is choking in a restaurant, and his wife calls for a doctor. A doctor steps forward from an adjacent table and applies the Heimlich Maneuver. In a few seconds, the man is breathing normally. The man thanks the doctor and says, "What do I owe you?"
> The doctor says, "Half what you would have paid me when you were choking to death."

The value of what you do invariably seems higher before you do it, than after. Get an agreement on costs and fees from your client **before** starting the work, because afterwards, you have limited bargaining power.

Also, an agreement beforehand will tell you if the client has a small budget and wants a basic job before you start work and not after.

PROJECT MANAGEMENT ONE-LINERS

q. Who owns the float? ...the contingency? ...any excess budget?

You do! Whether the float is schedule days or budget dollars, someone eventually will ask for it. As the Project Manager, you decide who deserves it.

r. What will get you fired, and what will cause your client never to retain your firm for another assignment?

List all the reasons to deal with each potentiality. And because listing the reason doesn't mean it won't happen, you must proactively address each reason by taking an action. For example, say you determine one reason is a subconsultant will fail to do its job properly and timely. You then could create a process of checking that sub's work in the sub's office at least weekly by assigning that task to a staff member who is told their performance will be evaluated on whether or not the sub is successful. However, remember in the end, it's really your performance as Project Manager that's evaluated on whether the sub performs well or not.

s. Problems go up geometrically with each interface.

The more offices and firms involved with a project, the more likely something goes wrong. For example, someone makes a change and forgets to advise every other party. As Project Manager, use systems to keep everyone involved with each step to reduce the likelihood things fall through the cracks. Systems include having an all-staff meeting at least monthly and making sure everyone who should be copied on key documents is copied. And because posting something on a website does not mean everybody knows they are expected to read it, you must tell people <u>each time</u> they are to read what is posted.

A virtual office (where much of the work is farmed out among several offices) is an extreme example of interface risks. Managing by teleconference works only up to a point. Make frequent visits to assess performance or be prepared to be disappointed every so often. Providing budget for these visits will be money well spent.

t. Some projects approach a Perfect Storm.

I reviewed eleven projects that had large financial losses and noticed six causes for margin erosion occurred repeatedly. If a project had more than three of those causes, we had a "perfect storm."

(1) <u>Price was too low</u>. In eight of the eleven projects, we "priced" the work too tightly. We either:

- Completely underestimated the scope which we were agreeing to provide;
- Left the scope so vague that the client could ask for anything and prove it was "in scope;"
- Shaved the price to win a price bidding competition.

In any case, we had no contingency to handle the unexpected. It was not that we overran a proper budget. Rather, the budget was inadequate before we began.

(2) <u>No recent project experience with the client</u>. In seven of the eleven projects, the client was one for whom we had never worked or had not worked recently. We didn't know the client's procedures and standards and had to learn on-the-job. We failed to build budget into such projects to cover the costs of learning.

(3) <u>Weak Technical Skills</u>. In seven of the eleven projects, the work was performed by staff lacking the technical skills to do it. In one case, an office without a bridge group signed an agreement to design a major bridge. To accomplish the work, they hired two bridge engineers and gave them the work to do before knowing the new hires' strengths and weaknesses. We accepted too many key staff of unknown skills on the project.

(4) <u>New Project Manager</u>. In six of the eleven projects, this was either the first project for the Project Managers or the first project of its type they managed. For example, we assigned a $5 million (fee) design project with four subcontractors to a Project Manager who had only managed $50,000 planning studies. It wasn't surprising that a new Project Manager didn't know to

budget for constant support as a mitigation strategy in the project risk mitigation plan.

(5) <u>Poor Control of Subcontractors</u>. In five of the eleven projects, the poor performance of a subcontractor caused us to incur significant extra costs when the Project Manager failed to proactively control the sub. In one instance, the sub had specialized expertise, and we failed to designate someone qualified to review the sub's work.

(6) <u>Overhead Charged to Project</u>. In three of the eleven projects, substantial costs which should have been charged to overhead (such as standby time) were charged to the project. Unfortunately, some Project Managers find it hard to resist when pressured by their supervisors to allow staff without work to charge time to the project. You must do everything you can to guard your project from being burdened by these overhead costs.

Assess your proposed and current projects to determine where you have to protect against any of the causes. If three of the above six causes for margin erosion apply to your project, you are in a high risk, danger zone. In fact, even one or two causes should be a red flag to you. You must be on the lookout for a problem arising when warnings are so obvious. Once you visualize the potential for a disaster, immediately take the steps to avoid one.

4. ORGANIZATION

a. There's no one-size-fits-all for a project organization structure.

What once was a successful organization structure may not work the next time. Change the structure as people rotate on and off the project, when the project moves into a different phase, or whenever the scope of services is revised over a project's life. Tailor your organization to take advantage of the strengths of the people on your team.

If everyone who was in a certain position seems to fail, consider whether the structure inhibits the holder from being successful. A change in structure then may be warranted, especially when you believe all who had the position were qualified.

A hierarchical (pyramid-shaped) organization can be a problem if the person at the top makes it clear that only he or she, the boss, can make a final decision. Before long, everyone else stops making a decision and starts passing all decisions up to the top. The micromanaging boss becomes a bottleneck.

Problems bubble-up quickly in a matrix organization, because a person in a matrix organization serves two masters. When that person receives conflicting instructions, it rings a bell that something is wrong, and then the organization's leader can step-in and deal promptly with the problem. However, as the leader, a Project Manager must be sensitive to the feelings of those uncomfortable in a matrix organization.

b. Rotating the tires doesn't make the engine any better.

Think of the engine as your team, and the tire positions as the organization structure. If things aren't going well, we often consider changing the structure. However, when the team isn't strong, changing the organization structure is unlikely to help very much.

c. A subconsultant shouldn't provide the Deputy Project Manager.

To most observers, a deputy fills in when the Project Manager is away. However, when we were the prime, I didn't want the Deputy Project Manager from a subconsultant committing me to do anything. Even if the deputy couldn't contractually commit me, I still would feel obligated to comply in order to keep the client happy.

For that reason, I avoided agreeing that the sub provides the Deputy Project Manager. I could still give an important position in the organization to a key participant from the sub; I just didn't designate the position Deputy Project Manager.

APPENDIX B

GOOD PRACTICES/LESSONS LEARNED

<u>Chapter 1   Civil Engineering Student at CCNY</u>

    1-1.   Include an executive summary at the front of a report.
    1-2.   Criticize the action, not the person.
    1-3.   It's more important to be courteous than prideful.
    1-4.   Learn in advance what you can about those who will be interviewing you, whether for an employment position or a business opportunity.
    1-5.   When reviewing the work of others, use key indicators to assess if they understand what they are doing.
    1-6.   No one is so important that they can't pitch in to help at the most elementary effort when necessary.

<u>Chapter 2   Junior Engineer</u>

    2-1.   Take action if anyone you work with appears biased. Promptly, deal with the situation if you're senior management. When you're not, speak with someone senior to have them address the circumstances.
    2-2.   When designing something, consider whether the contractor or manufacturer could bid and build the item and an inspector could determine if the work was constructed or manufactured as intended.

<u>Chapter 3   Highway Engineer</u>

    3-1.   Criticize constructively.
    3-2.   Before issuing a final report or plan, confirm what exists at the site is what's depicted in the report or plan.
    3-3.   Realize a casually conveyed statement is open to misinterpretation.
    3-4.   Dress appropriately.
    3-5   Speak up when you notice bias against others, rather than assuming someone else will do something.

Chapter 4 — Squad Leader

4-1. Begin planning for the next work effort while completing your current one, because your current work effort often is part of a project continuum.

4-2. Participate in social activities at work as it can be useful in many career-related ways.

4-3. Make people aware you will check everything to reinforce the fact they should do it right the first time, thus reducing the number of do-overs.

Chapter 5 — Project Administrator

5-1. Try to sniff out what's going wrong, as nothing goes perfectly. For example, determine who will overrun their budget, as it's likely someone is on track to do so. Likewise find out who will probably deliver a poor product and disappoint you.

5-2. Perform project administrative tasks at least once, to both appreciate and review the efforts of those who perform them.

Chapter 6 — Project Manager

6-1. Create a list of deliverables from the contract and check-off listed items as they are completed.

6-2. Read the entire contract monthly, including all the boilerplate terms and not just the technical scope of services, until you essentially memorize it. Also, read the chronological file monthly.

6-3. Appreciate the impacts of property acquisition and utility relocations on a project's cost and schedule, as they often are more than originally estimated.

6-4. Make the effort to learn the underlying elements to a problem so you will avoid giving useless answers that respond to the wrong base question.

6-5. Apply too much additional help rather than not enough when you're behind schedule and require more resources. Don't risk missing a due date by adding insufficient additional resources, when you probably can make the date with still more resources.

GOOD PRACTICES/LESSONS LEARNED

6-6. Require staff who attend conferences and seminars to prepare a written summary or hold an internal presentation to share new ideas and concepts they learned.

6-7. Create a new system or control when a mistake happens, to protect against that mistake happening again.

6-8. Keep your commitments; if you said you were going to deliver something by a certain date, deliver it. If that's not possible, warn your recipient when a deliverable date is going to be missed so the recipient can plan for the delay, say by rescheduling its internal review team. If you're going to be three days late, tell the recipient at least three days before the due date.

6-9. Be punctual to show respect for those with whom you're meeting.

6-10. Check every letter and e-mail for errors and typos, because a poorly written document reflects badly on the organization and the writer and implies the document's conclusions are suspect.

6-11. Write each letter and e-mail so it has only one topic. Be clear about the action you want the recipient to take (e.g., approve, respond with comments, send some information, etc.).

6-12. List every transmitted document and its unique title or document number in the transmittal letter, so you have proof of what you did.

6-13. Create individual action lists for those working on your project.

6-14. Constantly check to confirm staff are hard at work on your project. Asking if they are working isn't good enough as people often won't admit they are working on someone else's project. The best way to find out what is really happening is to periodically view the work products people had completed to date, such as working drawings or a draft of a report.

6-15. Collect articles you feel may be useful someday from newspapers, technical journals, etc. File the articles in such categories as speaking, writing, insurance, expert testimony, business development and quality.

6-16. Keep quiet if you are in over your head.

Chapter 7   Managing a Multi-Disciplinary Project

7-1. Understand that people from disparate groups often think differently and getting them together periodically is worthwhile,

244   THE ENGINEERING IS EASY

      if only for each to see that the other wants the same thing: A successful outcome.

7-2. Resist being pressured to reach a conclusion before you're ready.

7-3. Describe items in reports or plans that may appear flexible, but shouldn't be changed.

7-4. Look at a computer's output to see if it passes a sanity check, rather than blindly accepting the data.

Chapter 8   Managing Multiple Multi-Disciplinary Projects

8-1. Maintain a portion of a project's budget for contingencies as something unexpected invariably occurs, and you will require resources to resolve it.

8-2. Schedule weekly internal meetings of key project staff on active projects to share information and reduce surprises.

8-3. When negotiating, identify if the other party uses a devious approach, to avoid being at a disadvantage during negotiations.

8-4. Be frugal with a project's money.

8-5. Upon commencing an effort, set checkpoints and benchmarks in the process when the original approach should be reviewed for possible revision.

8-6. Confirm all results are free of partiality.

Chapter 9   Managing Projects for Private Sector Clients

9-1. Be wary of clients with a poor history of paying on time.

9-2. Fit into a client's comfort zone, including everything from dress code to work patterns.

9-3 Learn and focus on what's important to your client and your supervisor.

9-4. Seek opportunities to train all staff, especially those who are up-and-coming.

9-5. Be compassionate with staff with personal problems. Show you have empathy for them.

Chapter 10   Managing a Mega-Project in New York City

10-1. Ask for advice, as it is a sign of strength, not weakness.

GOOD PRACTICES/LESSONS LEARNED

10-2. Involve everyone in decision-making, including asking subordinates for advice and comments. Learn when and how to delegate.

10-3. Be aware that what works for small or simple projects may not work for large or complex ones.

10-4. Give fair warning to people of possible major changes to their work assignments.

10-5. Perform background checks of new hires.

10-6. Oversee and check the work of subconsultants and consultants.

10-7. Ask questions and be open to hear answers that disappoint you.

Chapter 11   More Project Management Assignments

11-1. Make the best of each project or posting offered to you; a positive attitude motivates those working with you.

11-2. Vet the capabilities of all staff assigned to your project to avoid struggling with unqualified staff. Use staff to take advantage of their strengths. Provide additional oversight when you're concerned they may be working on tasks where they are weak.

11-3. Be wary of offering the Deputy Project Manager position on your project to a subconsultant.

11-4. Think before speaking to avoid making an insensitive statement.

11-5 Make the effort to learn the local issues when pursuing a project as its Project Manager, rather than counting on others to know those issues on your behalf.

11-6 Realize the time it takes to reach a decision increases as the number of key stakeholders increases. Anticipate that different stakeholders could have different desired project outcomes.

11-7 Look for your next project rather than marking time on a project going nowhere.

Chapter 12   Managing an Asset Management Project in New York City

12-1. Develop a credible approach to validate subjective findings.

Chapter 13   Managing Operations

13-1. Tell your supervisor of problems so he/she first hears about it from you.

13-2. Seek to work with the best people you can to increase the likelihood of success.

13-3. Spend the time necessary to observe your staff's performance and help them until satisfied they are qualified. Accept that the fault is yours, not theirs, when you appoint someone unqualified to fill a slot and they fail. Avoid giving short shrift to those not easily or conveniently observed.

13-4. Consider changing the organizational structure when person after person fails in the same position, as the structure may be inhibiting success.

13-5. Spend sufficient time both serving and observing remote offices you oversee. Avoid becoming the supervisor in a large, well-equipped home office who doesn't appreciate the hardships faced by those in smaller field or branch offices.

13-6. Realize not everyone acts the same, as people act differently when they are evaluated differently.

13-7. Accept that when you wouldn't like filling in a new form you created, the form probably is more complicated and onerous than it needs to be.

13-8. Treat everyone with respect.

Chapter 14   Project Reviews

14-1. Determine a project's risks at the start, and budget for those risks accordingly.

14-2. Consider a change-in-command audit when key staff change.

14-3. Participate in project peer reviews as they are useful for both the reviewer and those on the project.

14-4. Control the potential for margin erosion before there is margin erosion.

14-5. Reach out to an independent expert when you need an impartial arbitrator to break a stalemate.

14-6. Perform a quality review before a submittal. If there isn't time to do a review beforehand, do one immediately after the submittal.

14-7. Hold periodic all-hands meetings to help maintain positive morale on your project.

14-8. Identify and deal with bad apples in your organization.

14-9. Confirm all policies are being followed, rather than assuming they are.

14-10. Realize over-confidence causes people to ignore potential risks.

GOOD PRACTICES/LESSONS LEARNED

14-11. Constantly emphasize a commitment to ethics, realizing there always will be temptation to act improperly.

14-12. Ask for help if you face a paradigm shift in your experiences.

Chapter 15   International Efforts

15-1. Start planning for major events as soon as practical.

Chapter 16   Managing a Design-Build Project in Turkey

16-1. Agree only to what's in your level of expertise.
16-2. Check the quality of all deliverables.
16-3. Acclimate to the culture where you are assigned to work.
16-4. Show hard-working staff you appreciate their efforts.
16-5. Confirm criteria and standards you've used previously apply before using them in a new environment.

Chapter 17   Managing a Project in Hong Kong

17-1. Look for opportunities to publish and present papers, especially if you can co-write one with a client or supervisor.
17-2. Learn the tax code where you will work, including its impact upon project costs.

Chapter 18   Managing a Program in Guam

18-1. Be prepared to describe what changed, such that you could reduce a previously submitted budget.
18-2. Accept that finding qualified personnel to staff a remote site typically is a lengthier process than contemplated. Realize quality suffers until sufficient, qualified staff are on board.
18-3. Avoid scheduling stays of a few days in a remote location as they are rarely cost effective.
18-4. Avoid talking negatively about someone to a third party without a compelling reason to do so.
18-5. Strive for alignment of scope, schedule and budget in a timely fashion, even when frequent scope changes make it challenging to do so.

248 THE ENGINEERING IS EASY

18-6. Understand the more offices and firms involved with a project, the more difficult coordination will be and, therefore, the more resources and time will be required to maintain quality.

18-7. Document, document. Make a written record of everything to be able to contest anyone who forgets or misinterprets what was decided.

Chapter 19   Managing a Public-Private Partnership Project in San Diego

19-1. Make sure everyone, especially those in the field, is safety conscious.

19-2. Allow for the sensitivity of trends when performing calculations.

Chapter 20   Miscellaneous Responsibilities

20-1 Accept extra assignments only when you're sure you can do both your current and the new assignments satisfactorily.

20-2. Be wary of relying on anyone you don't really know.

20-3. Seek ways to turn a paradigm shift into a marketing opportunity.

Chapter 21   Two Not So Successful Program Management Assignments

21-1. Have substantial schedule and cost contingencies on hardware-software development projects.

21-2. Advise your supervisor and the client as soon as you are aware something went wrong or something significant happened.

21-3. Act decisively on negative findings.

21-4. Use what-if meetings to strategize different potential approaches to achieve the desired outcome.

21-5. Put your concerns in writing to the client, even when you know they don't want you to do so.

21-6. Avail yourself of legal assistance when potentially facing legal action.

Chapter 22   Managing a Mega-Project in Houston

22-1. Be aware that many people are passionate about the office space assigned to them.

22-2. Anticipate and prepare in advance for your client's requests.

22-3. Hold periodic safety training sessions.

GOOD PRACTICES/LESSONS LEARNED

22-4. Treat risk management seriously.
22-5. Realize that estimates of final costs and time for completion of mega-projects often are too optimistic.
22-6. Consider using ranges for estimating a project's potential cost.

Chapter 23   Firm-Wide Director of Projects

23-1. Teach others to help maintain your own skills.
23-2 Avoid using sarcasm as it rarely works as you hope.

Chapter 24   External Efforts

24-1. Consider volunteering to provide community service as it's likely such service will improve your skills.
24-2. Be active in outside organizations to both serve your profession and develop new personal skills.
24-3. Treat competitor's personnel with respect.

Chapter 25   Proposal Management

25-1. Focus your efforts on pursuing clients who are most likely to select your firm for an assignment.

Chapter 26   Director on an International Engineering Company Board

26-1. Speak to a senior manager or use your organization's hotline to report someone who may be acting inappropriately.

Chapter 27   Reflections

27-1. Work harder than others.
27-2. Set aside your egotistical needs for the good of the organization.
27-3. Surround yourself with experts in disciplines that aren't your strength.
27-4. Increase the number of deputies and key administrative staff as a project's size increases.
27-5. Allow for the fact it may take a long time to learn a new client's or supervisor's procedures and how fairly they will treat you.

27-6. Document all decisions in case your client counterpart changes, as it's unlikely the new one will agree with every decision the previous one made.

27-7. Learn the strengths and weaknesses of new staff, so you are less likely to be disappointed in their performance.

27-8. Vet key staff before they are assigned to work with you, so you are less likely to wind up with staff who don't fulfill your requirements.

27-9. Anticipate that inexperienced staff won't perform to the same level as experienced. Allow for more time and do-overs when using inexperienced staff.

27-10. Make sure you have sufficient additional budget and duration when staff are scattered in different offices, as that situation is difficult to control.

27-11. Start developing your successor from Day 1.

27-12. Be willing to consider relocation as an option for your next assignment.

27-13. Make the effort to learn what skills others have, rather than prejudging others' talents as compared with your own.

ABOUT THE AUTHOR

Bruce Podwal received a B.C.E. from City College of New York (CCNY) and an M.S.C.E. from Polytechnic Institute of Brooklyn. A licensed Professional Engineer in eight states, he managed major engineering projects including the $10 billion Westway project in NYC, $200 million Guam Island-Wide Program, $2 billion Ankara-Gerede Motorway in Turkey, Hong Kong's $400 million Central Kowloon Route, and Houston's $3 billion Katy Freeway. He provided technical expertise on the Panama Canal Expansion, the U.S. Strategic Petroleum Reserve, and Dubai's Palm Island. Podwal served on the board of Parsons Brinckerhoff, a 12,000-person engineering firm, where he was manager for the region from New York to Virginia and CEO of several of its subsidiaries. He served on the South Bay Expressway LLC board as the representative of the U.S. DOT, a partial owner of this privatized San Diego toll road.

A Fellow of ASCE, Podwal was elected a Director to ASCE's National Board. He served on ASCE's Committee on Advancing the Profession and the editorial board of its "Leadership and Management in Engineering" journal and chaired its International Activities Committee and Task Committee to Study Professional Liability & Risks/Claims Management. He was President of ASCE's NYC-based Metropolitan Section and received its Thomas C. Kavanagh Service Award.

Podwal co-authored "Highway Engineering," in McGraw-Hill's *Standard Handbook for Civil Engineers*, 2nd and 3rd editions. He authored and presented numerous papers including "Consortium Teams: How to Make Them Work Multinational," at ASCE's Annual Conference and chaired the California Engineering Foundation's "Public-Private Partnerships in Transportation" conference. Podwal presented a two-week seminar on the design of highways, toll roads, tunnels and bridges at the Beijing Municipal Engineering Design and Research Institute and guest lectured at the University of Houston and Turkey's Middle East Technical University.

Podwal chaired the CCNY Dept. of Civil Engineering Advisory Board and presented "Project Management Challenges of Major International Engineering Projects," at the CCNY seminar series named in his honor. He was elected a CCNY Chapter Honor Member of Chi Epsilon (the civil engineering honor society) and received the CCNY Engineering School Alumni's Career Achievement Award.

www.ingramcontent.com/pod-product-compliance
Lightning Source LLC
Chambersburg PA
CBHW021939290426
44108CB00012B/894